KT-398-439

MANAGING THE MODERN HERBARIUM

An interdisciplinary approach

Deborah A. Metsger and Sheila C. Byers, Editors

"This resource book is the most comprehensive account of the care and management of plant and fungal collections to be published in the last decade. The subjects covered are extremely broad ranging from broad questions of preventive conservation to the hottest topics of destructive sampling in molecular systematics research and Integrated Pest Management.

Some chapters read like the Consumer Reports for herbaria, for example evaluating adhesives, making recommendations, and listing the names and addresses of suppliers. Experienced curators share their practical knowledge of constructing and moving million-specimen collections with valuable insights into their mistakes and successes.

Career opportunities are covered including lists of schools with graduate programs in collections management. International in scope, this book is recommended for those involved in managing plant and fungal herbaria as well as everyone interested in documenting the earth's plant and fungal biodiversity."

Dr. Amy Rossman, Director, U.S. National Fungus Collections, Beltsville Agricultural Research Center, Maryland

DR004429

CORNWALL COLLEGE

"Herbaria, collections of documented, preserved plant and fungal specimens, are one of the oldest and most basic elements of botanical systematics, and yet they are becoming increasingly important as years go by. Why are physical voucher specimens needed? Why isn't the literature sufficient? Human-caused modifications to the world's flora (e.g., habitat destruction leading to extirpation, introduction of weeds, and climate change) increasingly leave the herbarium as the only verifiable record of baseline conditions.

New technologies (e.g., morphometrics, DNA sequencing) allow the extraction of new kinds of data, requiring access to physical specimens. Nomenclatorial changes brought about by improved understanding of relationships make reference to physical vouchers necessary to interpret the meaning of the past literature.

This book covers a diverse set of issues ranging from pragmatic specimen preservation, to information science, to a very thoughtful and sensible set of recommendations for destructive sampling for molecular studies.

Plant systematists these days require so much training in new areas of research but still, perhaps more than ever, need a grounding in herbarium curation and management; this book should thus be required reading for all graduate students in the field of plant systematics. It sets a new standard."

Dr. Brent D. Mishler, Director, University and Jepson Herbaria,
and Professor, Department of Integrative Biology,
University of California, Berkeley

MANAGING THE MODERN HERBARIUM

An inter-disciplinary approach

Deborah A. Metsger
Sheila C. Byers
Editors

SPNHC

ROM

© 1999 Society for the Preservation of Natural History Collections, Washington, DC
SPNHC Managing Editor: Kate Shaw

This is contribution number 53 from the Centre for Biodiversity and Conservation Biology of the Royal Ontario Museum.

All rights reserved. No part of this publication may be reproduced without express written permission of the publisher, except in the case of brief quotations embodied in critical articles or reviews.

ISBN 0-9635476-2-3
LOC 99-64548

Library of Congress Cataloguing-in-Publication Data
Managing the Modern Herbarium—
An inter-disciplinary approach
(1999: Society for the Preservation of Natural History Collections, PO Box 797, Washington, DC 20044-0797, USA)
Includes bibliographic references

1. Metsger, Deborah A., 1956–. 2. Byers, Sheila C., 1950–.
3. Society for the Preservation of Natural History Collections.

Production and Distribution:

ELTON-WOLF PUBLISHING
Suite 212 - 1656 Duranleau St.
Granville Island
Vancouver, BC, Canada
V6H 3S4

Text Design: Fiona Raven
Cover Design: Jan Perrier

First printing June 1999

Printed in Canada

Note: The covers of this book are 10 point Cornwall. The text is printed on Weyerhaeuser Offset Opaque paper in 60 lb weight. This is a permanent and durable paper, manufactured in compliance with the standards set for alkaline papers by American National Standards, Inc. and the American Society for Testing and Materials. These papers are listed as "acid-free."

MANAGING THE MODERN HERBARIUM

An inter-disciplinary approach

Deborah A. Metsger
Sheila C. Byers
Editors

Published by:
Society for the Preservation of Natural History
Collections

as a joint project with:
The Royal Ontario Museum, Centre for
Biodiversity and Conservation Biology

1999

PHOTO CREDITS

Note: copyright © rests with photographers unless indicated otherwise.

Front Cover

Clockwise from top right:

Black cherry (*Prunus serotina*) - M. Ferguson, FPSA[1],
  © Royal Ontario Museum
Chanterelle (*Cantharellus infundiboliformis*) - M. Ferguson, FPSA,
  © Royal Ontario Museum
Pressing aquatic plants - J. Warren, © Royal Ontario Museum
Mounted specimens piled while glue dries - T. A. Dickinson,
  © Royal Ontario Museum

Back Cover

Clockwise from top right:

Riparian marsh, Bloodvein River, Ontario - D. A. Metsger,
  © Royal Ontario Museum
Mounting specimens - W. Robertson, © Royal Ontario Museum
Herbarium specimen of sugar maple (*Acer saccharum*) - B. Boyle,
  © Royal Ontario Museum
Herbarium cabinet and specimens in use - W. Robertson,
  © Royal Ontario Museum
Gynoecium of *Physocarpus opulifolius* in cross section - R. C. Evans
  © University of Toronto
Composite of Southern blot autoradiographs - R. C. Evans
  © University of Toronto
Tiger swallowtail butterfly (*Pterourus glaucus*) on a day lily
  (*Hemerocallis lilioasphodelus*) - Mary Ferguson, FPSA,
  © Royal Ontario Museum

---

[1] FPSA: Fellow of the Photographic Society of America

SOCIETY FOR THE PRESERVATION OF NATURAL HISTORY COLLECTIONS

SPNHC   The Society for the Preservation of Natural History Collections (SPNHC) was founded in 1985. It is an international, multidisciplinary organization comprised of individuals from the areas of anthropology, botany, geology, palaeontology, and zoology, as well as others who are interested in the development and preservation of natural history collections. Natural history collections consist of specimens and supporting documentation such as audio-visual materials, labels, library materials, field data, and similar archives. Preservation refers to any direct or indirect activity providing continued, and improved care of these collections and their documentation. The Society encourages the participation of individuals involved with all aspects of natural history collections.

ROYAL ONTARIO MUSEUM

ROM   Founded in 1912, the Royal Ontario Museum is one of Canada's preeminent cultural and scientific institutions. The ROM has attained international recognition for its significant collections of art, archaeology, and natural history, as well as for its extensive field activity. Its collections and research activities are complemented by an array of programs in exhibition, education, and public outreach. All of the ROM's natural history collections originated at the University of Toronto, of which the Museum was a part until 1968. The ROM's Centre for Biodiversity and Conservation Biology was formed in 1996 as a result of the amalgamation of the seven life science disciplines: botany (including mycology), entomology, herpetology, ichthyology, invertebrate zoology, mammalogy, and ornithology. The Centre is currently the custodian of collections of several million specimens.

This book is dedicated to

PRESERVING THE RECORD OF NATURE

THROUGH COUNTLESS AGES

*As inscribed on the stone pillars
at the entrance to
the Royal Ontario Museum*

CONTENTS

FOREWORD

Underlying the conception, organization, and realization of the 1995 Society for the Preservation of Natural History Collections (SPNHC) Education and Training Committee workshop "Managing the Modern Herbarium" is a conviction about the primacy of herbarium specimens as documentation and resource in botanical systematics, floristics, and vegetation studies. Although imperfect in this role — herbarium specimens are limited in their ability to capture and preserve all aspects of plant diversity (Baum, 1996; Heywood, 1996) — they are in many cases all the tangible evidence available on which to base taxonomic judgments, accounts of floras, and analyses of vegetation processes. Nevertheless, when their limitations are understood and respected, herbarium specimens can be important tools for testing hypotheses concerning the biology of plant taxa. For example, integrating field observations of distribution patterns in vascular plants of fragmented northern hardwood stands in central Ontario with information from the literature led to using herbarium specimens to categorize the dispersal mode of these plants. This proved to be particularly informative for ant-dispersed taxa since only a small fraction of these had previously been assessed (W. Draper, University of Toronto, pers. comm.). In this case it was not only herbarium specimens themselves but also a vouchered reference collection of seeds and fruits (Montgomery, 1977) that proved to be of value in developing morphologically-based hypotheses for testing in the field.

As plant systematics (and comparative biology generally) has come to look at wider and wider arrays of evidence, herbaria have also accumulated a variety of ancillary collections. The potential uses of herbarium specimens and of these ancillary collections are not always foreseen by collectors and curators. For instance, many herbaria have impressive collections of wood, fruit, and seed samples that, like herbarium specimens themselves, can be of immense value, not only in the primary research context in which they were assembled, but also in a variety of public service areas (e.g., as authentic material with which to compare specimens submitted for identification by medical, law enforcement, or other personnel). Pharmaceutical studies may result in herbaria amassing and possibly storing bulk collections of bark, roots, leaves, fruits, and seeds (Miller, 1996). Herbaria may also come to house alcoholic collections of mature or developmental stages of plant reproductive and vegetative structures that can take researchers far beyond the level of detail provided by rehydrated herbarium material.

With the emergence of molecular systematics, fostered by technological breakthroughs in extracting and analyzing initially minute amounts of plant DNA, two related issues with which herbaria (and systematics collections generally) have dealt in the past have once again become critically important: whether or not there is a herbarium voucher associated with any given DNA extract or nucleotide sequence and, if there is, whether this specimen is in fact the sole source of the tissue from which DNA has been (and can be) obtained for the taxon in question. As suggested by authors in the symposium on destructive sampling and molecular systematics, there is the potential for differences in scientific culture between molecular systematists and herbarium personnel that could lead to indifference on the part of the former to the need to associate any given DNA extract with a tangible, identifiable, and accessioned specimen. Similarly, issues associated with destructively sampling unique specimens may not always be understood sufficiently, or in the same way, by these two groups. It is thus of considerable importance that, in addition to the papers in Chapters 15–20 that review current practice in herbaria, entomological collections, and vertebrate collections, the workshop symposium on destructive sampling and molecular systematics also prepared recommendations for the use of museum collections in this context (Chapter 21).

In keeping with the pragmatic focus on specimens, their care, and their uses, it was altogether fitting that the workshop also included an opportunity for information exchange not only by means of research papers on herbarium design, archival materials, and collection management tools (Chapters 1-14) but also in the form of the Herbarium Information Bazaar. This evening activity provided a venue not only for the presentations documented in Part III, but also for vendors of herbarium and related equipment (e.g., specimen preparation and storage supplies, bar coding equipment). The workshop as a whole thus succeeded admirably in providing those attending with a broad range of information relevant to herbarium operations to take back to their collections and colleagues in order to better look after the specimens in their care. The present publication will make the same broad range of information accessible to a much wider audience.

T. A. Dickinson,
Curator, Vascular Plant Herbarium (TRT),
Centre for Biodiversity and Conservation Biology,
ROYAL ONTARIO MUSEUM

LITERATURE CITED

BAUM, B. R. 1996. Statistical adequacy of plant collections. Pp. 43-73 in T. F. Stuessy and S. H. Sohmer, editors. *Sampling the Green World*. Columbia University Press, New York, xvi+289.

HEYWOOD, V. H. 1996. A historical overview of documenting plant diversity: are there lessons for the future? Pp. 3-15 in T. F. Stuessy and S. H. Sohmer, editors. *Sampling the Green World*. Columbia University Press, New York, xvi+289.

MILLER, J. S. 1996. Collecting methodologies for plant samples for pharmaceutical research. Pp. 74-87 in T. F. Stuessy and S. H. Sohmer, editors. *Sampling the Green World*. Columbia University Press, New York, xvi+289.

MONTGOMERY, F. H. 1977. *Seeds and Fruits of Plants of Eastern Canada and Northeastern United States*. University of Toronto Press, Toronto, xi+232.

PREFACE

This book follows from a workshop entitled "Managing the Modern Herbarium" that was hosted jointly by the Botany Department of the Royal Ontario Museum (ROM) and the Education and Training Committee of the Society for the Preservation of Natural History Collections (SPNHC). The workshop was held at the University of Toronto, June 5–6, 1995 and was attended by 94 participants, including herbarium curators, collections managers, conservators, professors, and students from six countries.

The Royal Ontario Museum maintains two herbaria, the Vascular Plant Herbarium and the Cryptogamic Herbarium, TRT and TRTC respectively (Holmgren et al., 1990), both of which are now included in the ROM Centre for Biodiversity and Conservation Biology. Together, the two herbaria house approximately 800,000 specimens of vascular plants, modern and fossil pollen and seeds, fungi, lichenized fungi, and bryophytes. Over the past 15 years ROM botanists have worked to promote the value of specimen-based data, and to improve the preservation methods and storage materials used in these collections. Both the workshop and this book reflect their commitment to working toward the standardization of techniques and policies by facilitating communication and thereby reducing duplication of effort among all those concerned with the maintenance and use of botanical collections.

Through its committees the Society for the Preservation of Natural History Collections (SPNHC) serves as a liaison between the conservation/preservation community and the natural science disciplines. As part of their purview, the SPNHC Education and Training Committee offers practical training workshops at each annual meeting. "Managing the Modern Herbarium" was the first such workshop that was discipline-specific rather than topic-oriented.

Following extensive discussions with many members of the international botanical and mycological community the workshop was organized as three separate topical sessions:
• Preventive Conservation in the Herbarium was a one day symposium which drew on the expertise of conservation professionals and engineers to focus on the principles of preventive conservation that relate to the long-term care of botanical and

mycological collections. The papers resulting from these presentations are found in Part I of this volume.

• Contemporary Issues Facing Herbaria occupied the second day of the workshop. Papers from this session are presented in Part II of this book. An in-depth presentation and discussion on the use of bar coding in collections started the day. This was followed by a symposium, the goal of which was to provide a forum for dialogue between herbarium curatorial staff and molecular systematists while exploring four different disciplinary perspectives — vascular plants, fungi, insects, and vertebrates — on destructive sampling of museum material for use in molecular systematics. The symposium resulted in a position paper (Chapter 21) that is offered to the systematics community at large to assist in the formulation of protocols and policies for destructive sampling.

• The Herbarium Information Bazaar was designed to encourage transfer of techniques and technologies among institutions; to provide opportunities for demonstration and experimentation; and to promote dialogue amongst botanists, conservators, and participating suppliers. Abstracts or short papers developed from some of these presentations have been included in Part III of this book.

Effective communication between disciplines, or even between institutions or regions, is often hindered by different unspoken assumptions or priorities, or by differences in terminology. In this volume we have attempted to standardize symbols and potentially confusing terminology between papers. The acronyms used throughout the book to identify herbaria are those officially recognized in *Index Herbariorum* (Holmgren et al., 1990). The following definitions have been taken from SPNHC (1994):

collection management: the responsibility and function of an institution that fosters the preservation, accessibility, and utility of their collections and associated data. The management process involves responsibilities for recommending and implementing policy with respect to: specimen acquisition, collection growth, and deaccessioning; planning and establishing collection priorities; obtaining, allocating, and managing resources; and coordinating collection processes with the needs of curation, preservation, and specimen use. These responsibilities may be shared by collection managers, subject specialists, curators, and other institutional administrators.

conservation: the science of stabilizing a specimen, artifact, or collection by retarding those factors that contribute to its deterioration by physical and chemical means. This involves activities such as preventive conservation, examination, documentation, treatment, research, and education.

preservation: often used synonymously with conservation; in most conservation disciplines, preservation involves preventive measures, such as correcting adverse environmental conditions, and maintenance procedures; in natural science conservation, preservation also includes treatments carried out initially to prepare specimens.

preventive conservation: a facet of conservation that involves taking steps to prevent deterioration and damage to collections to ensure their long-term preservation; includes such activities as risk assessment, development, and implementation of guidelines for continuing use and care, appropriate environmental conditions for storage and exhibition, and proper procedures for handling, packing, transport, and use.

We hope that this volume will serve as a valuable resource and guidebook for all herbarium professionals in their efforts to customize herbarium facilities, operations, and policies to meet the specific needs and circumstances of individual collections as they move into the 21st century.

Deborah A. Metsger and Sheila C. Byers
October 1998

LITERATURE CITED

Holmgren, P. K., N. H. Holmgren, and L. C. Barnett, editors. 1990. *Index Herbariorum.* New York Botanical Garden, New York, USA, 693 pp.

[SPNHC] Society for the Preservation of Natural History Collections. 1994. Guidelines for the care of natural history collections. Sessional Committee on Common Philosophies and Objectives. Collection Forum 10(1):32-40.

ACKNOWLEDGMENTS

Managing the Modern Herbarium was made possible by the generous contribution and support of many people. Ann Pinzl sowed the seed for the workshop through her 1991 survey of collections management practices in herbaria. The initial proposal was based on discussions with David Brunner, Tim Dickinson, Mike Donoghue, Mark Engstrom, Deb Lewis, John McNeill, Bob Murphy, Amy Rossman, Tod Stuessy, Janet Waddington, and Emily Wood. Additional suggestions for the content and format of the workshop were provided by the caucus of botanists and mycologists who participated in the initial brain-storming session at the joint annual meeting of the Association of Systematics Collections (ASC) and the Society for the Preservation of Natural History Collections (SPNHC) held at the Missouri Botanical Garden, May 1994.

We thank Dr. Verna Higgins, past Chair, Department of Botany, University of Toronto, and her staff for providing facilities and services for the workshop. Numerous Royal Ontario Museum staff and students committed time and energy to the workshop organizing committee and to the smooth operation of the workshop itself. The Herbarium Information Bazaar was master-minded by Deb Lewis and Ann Pinzl who devoted many hours to organizing this innovative hands-on approach to sharing information. Finally all the workshop contributors and participants are to be thanked for two days of stimulating presentations, discussions, and recommendations which provided the substance for this book.

We are indebted to the contributing authors for their efforts and patience throughout the editorial process. Barbara Moore and Emily Wood, in their capacity as section editors, ably coordinated the peer review process. Our thanks to them and to all of the reviewers. The combined knowledge, editorial experience, and support of Kate Shaw and John Simmons, present and past Managing Editors, SPNHC, Janet Waddington, and Tim Dickinson have guided us through the ups and downs of the editorial process. Jenny Bull, our copy editor, has helped produce a book that would be accessible to an audience comprising several disciplines. We recognize Jo Blackmore of Elton-Wolf Publishing for her patient guidance through the publication process and for her creative marketing ideas. We also thank SPNHC for their financial commitment to the realization of this book.

Last but not least, we wish to thank our co-workers and our respective families, Jeff, Laura, and Peter Warren, and Ray Hryciuk for putting up with, and not giving up on us for the last four years!

Funding for the workshop and this volume was provided by grants from: The Museums Assistance Program of the Department of Canadian Heritage; The Bay Foundation; The Royal Ontario Museum Foundation Endowment Fund; and Huntington T. Block Insurance Agency Inc. We wish to thank SPNHC Canada and the ROM Centre for Biodiversity and Conservation Biology for managing these funds.

CONTRIBUTORS

KERRY BARRINGER
Brooklyn Botanic Garden
1000 Washington Avenue
Brooklyn, NY 11225–1099, USA
kerrybarringer@bbg.org

SHEILA C. BYERS
1024 West 7th Avenue, Suite 202
Vancouver, BC V6H 1B3, Canada
scbyers@intouch.bc.ca

TIMOTHY A. DICKINSON
Centre for Biodiversity and
Conservation Biology
Royal Ontario Museum
100 Queen's Park
Toronto, ON M5S 2C6, Canada
timd@rom.on.ca

MICHAEL J. DONOGHUE
Harvard University Herbaria
22 Divinity Avenue
Cambridge, MA 02138, USA
mdonaghu@oeb.harvard.edu

JANE L. DOWN
Canadian Conservation Institute
1030 Innes Road
Ottawa, ON K1A 0M5, Canada
jane_down@pch.gc.ca

MARK D. ENGSTROM
Centre for Biodiversity and
Conservation Biology
Royal Ontario Museum
100 Queen's Park
Toronto, ON M5S 2C6, Canada
marke@rom.on.ca

TORSTEN ERIKSSON
The Bergius Foundation
Royal Swedish Academy of Sciences
Box 50017
104 05 Stockholm, Sweden
torsten@bergianska.se

BARBARA ERTTER
University and Jepson Herbaria
University of California
1001 VLSB #2465
Berkeley, CA 94720-2465, USA
ertter@uclink4.berkeley.edu

JULIA FENN
Conservation Department
Royal Ontario Museum
100 Queen's Park
Toronto, ON M5S 2C6, Canada
juliaf@rom.on.ca

JUDY GIBSON
Botany Department
San Diego Natural History Museum
PO Box 121390
San Diego, CA 92112-1390, USA
jgibson@cts.com

OLIVER HADDRATH
Centre for Biodiversity and
Conservation Biology
Royal Ontario Museum
100 Queen's Park
Toronto, ON M5S 2C6, Canada
oliverh@rom.on.ca

RUTH HERZBERG
Division of Parks, Recreation and
Preserves
Iowa Department of Natural Resources
Wallace State Office Building
Des Moines, IA 50319, USA

GREGORY J. HILL
Archives Preservation Division
National Archives of Canada
Gatineau Preservation Centre
625, boul. du Carrefour
Gatineau, PQ J8T 8L8, Canada
ghill@archives.ca

CELIA ISON
Research Collections
University of Nebraska State Museum
W436 Nebraska Hall
Lincoln, NE 68588-0514, USA

ROBERT K. JANSEN
Department of Botany and Institute of
Cellular and Molecular Biology
University of Texas
Austin, TX 78713, USA
rjansen@utxvms.cc.utexas.edu

KAREN L. JOHNSON
Natural History Division
Manitoba Museum of Man and
Nature
190 Rupert Avenue
Winnipeg, MN R3B 0N2, Canada
kjohnson@mbnet.mb.ca

HYI-GYUNG KIM
Laboratory of Molecular Systematics
Smithsonian Institution
4210 Silver Hill Road
Suitland, MD 20746, USA
hyigyung@onyx.si.edu

JANIS KLAPECKI
Natural History Division
Manitoba Museum of Man and
Nature
190 Rupert Avenue
Winnipeg, MN R3B 0N2, Canada
jklapeck@mbnet.mb.ca

DEBORAH LEWIS
Department of Botany
Iowa State University
353 Bessey Hall
Ames, IA 50011-1020, USA
dlewis@iastate.edu

DENNIS J. LOOCKERMAN
CODIS – Section of the Crime
Laboratory Service
MSC 0461
5805 N. Lamar
PO Box 4143
Austin, TX 78765-4143, USA

WILLIAM P. LULL
Garrison/Lull Inc.
Box 337
Princeton Junction, NJ 08550, USA
gli@theoffice.net

DEBORAH A. METSGER
Centre for Biodiversity and
Conservation Biology
Royal Ontario Museum
100 Queen's Park
Toronto, ON M5S 2C6, Canada
debm@rom.on.ca

JAMES S. MILLER
Applied Research Department
Missouri Botanical Garden
PO Box 299
St. Louis, MO 63166-0299, USA
james.miller@mobot.org

R. RICHARD MONK
Museum of Texas Tech University
Box 43191
Lubbock, TX 79409–3191, USA
richard@packrat.musm.ttu.edu

BARBARA MOORE
Conservation and Collections
Peabody Museum of Natural History
Yale University
170 Whitney Avenue, PO Box 208118
New Haven, CT 06520-8118, USA
barbara.moore@yale.edu

SIMON MOORE
Hampshire County Council Museums
Service
Chilcomb House, Chilcomb Lane
Winchester, Hampshire SO23 8RD, UK
MUSMSM@hantsnet.hants.gov.uk

GREGORY M. MUELLER
Department of Botany
The Field Museum
Chicago, IL 60605-2496, USA
gmueller@fmnh.org

ROBERT W. MURPHY
Centre for Biodiversity and
Conservation Biology
Royal Ontario Museum
100 Queen's Park
Toronto, ON M5S 2C6, Canada
bobm@rom.on.ca

ANN PINZL
Nevada State Museum
600 North Carson Street
Carson City, NV 89701, USA
apinzl@lahontan.clan.lib.nv.us

RICHARD K. RABELER
University of Michigan Herbarium
North University Building
1205 North University Avenue
Ann Arbor, MI 48109-1057, USA
rabeler@umich.edu

LINDA RADER
Research Collections
University of Nebraska State Museum
W436 Nebraska Hall
Lincoln, NE 68588-0514, USA
lrader@unlserve.unl.edu

CAROLYN L. ROSE
Department of Anthropology
National Museum of Natural History
Smithsonian Institution
Washington, DC 20560-0166, USA
rose.carolyn@nmnh.si.edu

GEORGE F. RUSSELL
Department of Botany, MRC 166
National Museum of Natural History
Smithsonian Institution
Washington, DC 20560-0166, USA
russellr@nmnh.si.edu

MICHAEL J. SHCHEPANEK
Botany Section
Canadian Museum of Nature
PO Box 3443, Station D
Ottawa, ON K1P 6P4, Canada
mshchepane@mus-nature.ca

S. LLYN SHARP
Virginia Tech Museum of Natural
History Dept. of Biology
Virginia Polytechnic Institute and
State University
Blacksburg, VA 24061-0406, USA
llyn@vt.edu

TOM J. K. STRANG
Canadian Conservation Institute
1030 Innes Road
Ottawa, ON K1A 0M5, Canada
Tom_Strang@pch.gc.ca

JOHN TOWNSEND
RR # 2, PO Box 596A
Cobleskill, NY 12043, USA
townseje@cobleskill.edu

JAMES B. WHITFIELD
Department of Entomology
321 Agriculture Building
University of Arkansas
Fayetteville, AR 72701, USA
jwhitfie@comp.uark.edu

THOMAS F. WIEBOLDT
Virginia Tech Museum of Natural
History Dept. of Biology
Virginia Polytechnic Institute and
State University
Blacksburg, VA 24061-0406, USA

STEPHEN L. WILLIAMS
Strecker Museum Complex
Baylor University, PO Box 97154
Waco, TX 76798–7154, USA
steve_williams@baylor.edu

EMILY W. WOOD
Harvard University Herbaria
22 Divinity Avenue
Cambridge, MA 02138, USA
ewood@oeb.harvard.edu

Introduction

Deborah A. Metsger and Sheila C. Byers

THE CLOSE OF THE 20TH CENTURY –
AN ERA OF CHANGE

Over the last two decades of the 20th century the environment in which herbaria and other natural history collections exist has undergone significant changes. These changes have placed a number of pressures on all those who manage collections. *Managing the Modern Herbarium* speaks to these pressures by presenting a broad range of perspectives on herbarium operations and related topics. The ultimate goal of this volume is to foster communication between herbarium personnel, researchers, conservators, and members of other disciplines in order to more effectively manage and preserve systematic collections of plants and fungi for the future.

In this era of change social, political, and economic shifts have had a profound effect on natural history museums and other cultural and scientific institutions (Rose et al., 1993; Janes, 1995; Cato et al., 1996; Cholewa, 1997; AAM, 1998; Butler et al., 1998). Institutions are no longer assured of government funding and public support and so must increasingly strive for self-sufficiency (Butler et al., 1998). The public focus, or outer museum function as described by Humphrey (1991), has become important to the survival of all cultural and scientific institutions, even those who previously enjoyed a research-centred existence (Cholewa, 1997).

At the same time, the world is threatened by catastrophic losses of natural diversity (Black, 1993). Natural diversity cannot be managed or conserved without fundamental knowledge of living organisms and the ecosystems in which they live. Since natural history specimens provide physical vouchers of those living organisms (Boom, 1996), the systematics community at large is in a strategic position to respond to the natural diversity crisis by documenting and preserving information about the world's biota before it is irreparably lost (Alberch, 1993; Anonymous, 1994; Stuessy, 1996).

The task of inventorying and documenting natural or bio-diversity has placed emphasis on the need to coordinate research efforts between institutions and to standardize documentation procedures on regional, national, and global scales (Anonymous, 1994; Butler et al., 1998). Recent advances in information technology have increased and will continue to increase our capacity to collect and store data of all kinds (Stuessy, 1996). Systematic data can now be made accessible to a larger, more broadly defined audience via the Internet, and so needs to be targeted accordingly. Some even predict that by the year 2050 every specimen in the world, and its associated data and images, will be able to be accessed by researchers from the World Wide Web (Stuessy, 1996; Butler et al., 1998; Russell, Chapter 14).

Academic research in systematics is also in the midst of a revolution, with every aspect of traditional thought and practice being reconsidered (Davis, 1995). There has been a significant shift from the predominant use of morphological evidence to greater dependence on molecular evidence to determine evolutionary relationships. The cross-application of data between disciplines is now considered critical to understanding biological processes (Anonymous, 1994; Boom, 1996; Stern and Eriksson, 1996). Correspondingly, the very concept of what constitutes a good herbarium specimen, for instance, has expanded to include material that reflects the overall biology of the organism: diseased plants; those with evidence of animal interactions; morphological or ecological variants; etc. The inclusion of these materials in herbaria will increase the potential for ecologists, physiologists, and scientists from many other branches of biology to use data from herbarium specimens in their research (Stern and Eriksson, 1996).

Materials science has also experienced substantial changes in recent decades. The focus of conservation activity has shifted from an emphasis on treating and documenting individual objects to preventing deterioration (Rose, 1991). In the last fifteen years there has been a monumental increase in the number of studies and conferences dealing with natural history collections and their preservation (Faber, 1983; Waddington and Rudkin, 1986; Rose and Torres, 1992; Duckworth et al., 1993; Rose et al., 1993; Rose et al., 1995). Through these initiatives the cost-effectiveness and long-term benefits of a preventive conservation approach to collections care have come to be more widely accepted (Rose and Hawks, 1995), often challenging or displacing traditional collections practices and procedures.

Together these changes have increased the pressure on individual collections and on the people who manage them to make specimens available and relevant to a broader spectrum of users and, at the same time, to preserve the integrity of the specimens for future, as yet unknown research uses. Creating a balance between the use, management, and conservation of collections is a significant challenge (Williams and Cato, 1995). To meet this challenge and to address the impacts of these pressures on modern herbaria one must first understand the nature of these collections.

THE MODERN HERBARIUM IN CONTEXT

The techniques and procedures followed in herbaria often differ from other natural history collections. This is not only because of basic differences between plants and animals and how they are handled and distributed as specimens (Anderson, 1996), but also because they represent five centuries of inherited tradition (Stearn, 1957; Ogilvie, 1985; Gunn, 1994; Heywood, 1996). Herbarium collections house diverse kinds of specimens — sometimes representing several kingdoms and many classes — each with different requirements for specimen preparation and different protocols for identification and research use. Further, because many herbaria are part of university departments or botanic gardens, they have tended to be dissociated from the mainstream natural history museum community (Cholewa, 1997). Correspondingly, the majority of papers written on various aspects of botanical collecting, specimen preparation, and herbarium conservation and management have been written by botanists and published in botanical journals or by botanical institutions (Savile, 1962; Hicks and Hicks, 1978; Bridson and Forman, 1992). Articles of relevance to botanical collections that are published, for instance, in the conservation literature (Gunn, 1994) or the agricultural literature (Strang, Chapter 4) are often unfamiliar to many botanists.

For all these reasons, the recent advances in preventive conservation have not had the same impact on the care and maintenance of botanical collections as they have had on other natural history collections. Similarly, because of innate differences in organismal types — the unitary organization and mobile nature of many animal taxa (especially vertebrates) as opposed to the metameric organization and stationary nature of plants and fungi — some zoological museums developed special collections for use in molecular systematics, such as frozen tissue collections (Engstrom et al., Chapter 19), nearly a decade earlier than analogous special collections became an issue for herbaria. The preparation of materials for these special collections also differs, for while alcohol is an acceptable preservative for animal DNA (Whitfield, Chapter 18; Engstrom et al., Chapter 19), it destroys DNA in plants and fungi (Jansen et al., Chapter 16; Mueller, Chapter 17; Miller, Chapter 20).

MANAGING THE MODERN HERBARIUM:
AN OVERVIEW

The workshop "Managing the Modern Herbarium," on which this book is based, provided an opportunity for dialogue in which to bridge gaps in disciplinary perspectives on contemporary collections issues and to focus the discussions on specific herbarium concerns. The resulting papers are presented in this volume in topical sections.

PART I: PREVENTIVE CONSERVATION
AND COLLECTIONS MANAGEMENT

The papers in Part I are devoted to issues of Preventive Conservation and Collections Management. Some papers detail the experiences of curators and collections managers, while in others, conservators and conservation scientists lend their expertise to common herbarium practices.

Balancing the need for long-term specimen preservation against the need to make specimens available to the scientific community may be especially daunting to a herbarium curator with expertise in systematics but without training in curatorial procedures and protocols (Cato, 1988; Cholewa, 1997). He or she may well ponder: Where do we begin? Who determines collections management priorities and how?

To address these questions, John Townsend (Chapter 1) opens Section I: Preventive Conservation, Principles and Practice by introducing preventive conservation "as a management tool designed to help prolong the useful life of the collection." He explains that preventive conservation should be addressed in the broad context of the institution responsible for the collection and the particular circumstances under which the institution operates. Townsend promotes the conservation assessment as a tool for identifying and evaluating specific conservation needs. He introduces the interacting agents of deterioration that can affect collections, and cautions against the use of packaged solutions to address conservation problems.

The use of packaged solutions is often tempting, especially for herbaria that lack conservation support within their institutions. Even in these instances however, the implementation of a comprehensive conservation program need not occur in a vacuum. Carolyn Rose (Chapter 2) provides a succinct introduction to and listing of the many resources that are now available to provide expertise on collections care. These range from bibliographic resources to consulting services, assessment agencies, discussion groups, and list servers.

One of the most critical elements of preventive conservation is integrated pest management (IPM). IPM takes a systematic, layered approach to the prevention, detection, and strategic control of pest infestation. Tom Strang (Chapter 3) champions his IPM approach, based on the Framework for Preservation of Museum Collections developed by the Canadian Conservation Institute (CCI, 1994; Rose and Hawks, 1995), by first comparing it to an alternative pest control system developed for a herbarium by E. D. Merrill in 1948. Strang provides a concise summary, with examples, of the elements of an effective IPM program and concludes with some insightful comments on their implementation and an introductory resource list for pest identification and pest management.

The shift away from the use of pesticides in herbarium collections has met with some resistance (Townsend, Chapter 1; Strang, Chapter 3) and the pros and cons of alternative methods, particularly thermal controls, have been subject to considerable debate in the botanical literature. In Chapter 4, Strang argues that the evaluation of the effectiveness or appropriateness of any methodology used in a collection requires some quantitative measure of success that is based on an informed, pre-determined criterion. To illustrate his point, he uses seed viability as a dosimeter of the effects of thermal pest control measures on the integrity of herbarium specimens.

With the basic principles of preventive conservation in hand, Section II: Elements of Herbarium Design is intended as a road map to achieving herbarium construction or renovation that successfully addresses the needs of collections and their users. Unlike the renovation of an office building, in which planning for new facilities is likely to be the responsibility of a design team, curators and collections managers must often do all the planning themselves, especially in smaller institutions. Unfortunately not everyone faced with this task will have the luxury of their own prior experience or that of others. In many cases lack of awareness of the environmental requirements for the long-term care of natural history specimens has resulted in renovations that either go half-way to solving problems or create new ones in the process.

Each of the papers presented here covers a different aspect of herbarium renovation and design. There is broad consensus that first-hand observation of similar facilities, with an eye to both their successes and failures, is a necessary precursor to designing new or renovated collections space. Further, renovation or construction is not an isolated event but an ongoing process during which the design must be constantly re-evaluated and checked. Finally, one must also be prepared to redesign the procedures used in a new or renovated facility.

In Chapter 5, William Lull and Barbara Moore use their combined knowledge of environmental systems and conservation principles to recommend basic standards and other considerations for the various environmental control elements of a herbarium construction. They place particular importance on centering a renovation or construction project around an identified set of environmental goals each tailored to the needs of the collections. They caution that design features should always be developed from these goals rather then adapting goals to the facilities post facto.

The remaining papers in this section provide case studies of actual herbarium renovations. Focussing on factors outside the expertise of architects, Barbara Errter (Chapter 6) provides a unique checklist to the basic elements of herbarium layout and design as they relate to achieving optimum functionality of operations. Richard Rabeler (Chapter 7) discusses the many factors to consider in compactor installation. He includes not only the physical and structural aspects of compactor choice but also functional tasks associated with achieving a desired specimen configuration and maximizing the use of available space. Kerry Barringer (Chapter 8) discusses how conservation and environmental assessments were used to plan the renovation of the herbarium at the Brooklyn Botanic Garden. He addresses the same environmental factors outlined by Lull and Moore and highlights the success and failure of their application in an actual herbarium installation. He also provides an assessment of the effectiveness of the renovations and of an IPM program after five years of operation. Finally, Barbara Ertter (Chapter 9) provides a case history of moving herbarium cases during various stages of the renovation of the herbaria at Berkeley.

In Section III: Preventive Conservation Approaches to Herbarium Materials and Methods, the focus shifts from the macroenvironment of the collection space to the microenvironment of the specimens themselves. Contemporary procedures for making, labelling, and organizing herbarium specimens differ little from the procedures used by Linnaeus (Philosophia botanica No. 11, 1751 as in Stearn, 1957):

1   Plants not to be damp when brought together
2   no *parts* to be laid aside
3   moderately spread out
4   not really bent
5   *with flowering and fruiting* organs present
6   *dried* between dried paper
7   *very quickly* with a slightly hot iron
8   with the *press* moderately compressed
9   *glued* down with fish-glue
10 *in folio* (i.e., between a cover made by a folded sheet) always preserved
11 *one* only to a page
12 the sheet of paper not bound
13 the *Genus* entered above
14 the *species* and history on the back
15 *members of the same genus* between "phylyra" (*lit.* inner bark of the linden tree, *hence* stout paper)
16 arranged according to a system of classification

These methods reflect what Linnaeus considered to be the optimum technique for preserving and displaying valuable taxonomic characters required for classification. It is unlikely that Linnaeus and his contemporaries chose their materials for their ability to stand up in perpetuity, although some of the materials have turned out to be long-lasting and may have even included insecticidal compounds (Stearn, 1957). Because we are now keenly aware of the long-term value of collections and because we mourn those type specimens that have been lost (Stearn, 1957; Anderson, 1996), custodians of contemporary collections pay much more attention to long-term preservation and correspondingly to the archival nature of the products that we use. In reality even archival product formulations rarely stay the same so collections staff need to be equipped to continually evaluate the suitability of products. It is often more important to avoid poor products than it is to select those that are of superior quality (Williams and Monk, Chapter 12).

The four papers presented here, written by conservators and conservation scientists, provide herbarium staff with tools and resources so that they may be both discerning consumers when purchasing supplies and responsible practitioners when working with specimens. These papers reflect the scenarios and concerns raised in a panel discussion that followed the original presentations, and the consensus that one must use common sense to achieve a balance between sound conservation practices and the realities of the existing collections.

Of all the natural science collections, herbaria are the greatest consumers of paper. Gregory Hill (Chapter 10) discusses the chemistry of paper and introduces standards and procedures that have been developed by the paper conservation community for the treatment, care, and storage of paper-based objects. Hill concludes that although paper conservators can provide advice to the herbarium community there is a need for continued dialogue and collaborative research to develop standards and guidelines that reflect actual specimen usage.

Jane Down (Chapter 11) places her discussion of adhesives in a herbarium context with a historical overview of adhesive use in the herbaria of the Canadian Museum of Nature, and assesses the relative condition of, and risks associated with, specimens from each era of adhesive use. She provides conservation guidelines for the selection and application of adhesives and emphasizes the importance of conservation criteria such as the reversibility of a process, a concept largely overlooked in herbaria.

The label that is mounted on a herbarium specimen is often the only permanent record of the plant, therefore the integrity of the label is critical to maintaining the value and utility of the specimen. Materials that are used to record information on herbarium labels or on any other hard-copy documents are termed documentation media. Stephen Williams and Richard Monk (Chapter 12) stress the importance of determining the quality of a documentation medium before it is selected for long-term use in a collection. They introduce the factors that can affect permanence of documentation media and provide a series of simple tests that can be used by collections staff to assess dry documentation media.

Off-gassing of corrosive vapours or biocides from collections storage supplies and equipment can compound the health and safety hazards associated with collections and pose a risk to the specimens themselves. Only a few plastics are considered sufficiently non-corrosive to use with museum artefacts and specimens. Julia

Fenn (Chapter 13) details the properties and qualities of seven plastics that are considered safe for use in museum collections and displays. She describes a number of undesirable properties for museum plastics and provides guidelines and simple tests for identifying plastic materials that are already in use in herbaria. Because herbarium specimens and cases are frequently saturated with a mosaic of biocides, she also provides current threshold values and health risks for some biocides that were historically used in herbaria.

PART II: CONTEMPORARY ISSUES FACING HERBARIA

Thus far this volume has focussed on how principles of preventive conservation, modern technologies, a fundamental awareness of the specimens, and common sense can be combined to improve collections care and preserve the integrity of the specimens. Part II: Contemporary Issues Facing Herbaria looks more particularly at specimen usage in light of recent advances in information technology and in approaches to systematic research — making specimens available and relevant to a broader spectrum of users while preserving the integrity of the specimens.

As more and more herbaria have computerized their specimen data, the need to track specimens and specimen information has increased. Many collections have employed bar codes as a means of accomplishing this. Based on his experience at the US National Herbarium, George Russell (Chapter 14) de-mystifies the bar coding process and associated terminology. He provides useful considerations and guidelines for the selection of a bar coding system, the setup of the codes, and the choice of labels and adhesives. Russell suggests that in future it will be necessary to develop a community strategy for creating a system of unique bar code values and to agree upon a community authority for implementing that strategy.

The destructive sampling of specimens is not a new phenomenon in botany and mycology (Lawrence, 1951; Bridson and Forman, 1992; Wood et al., Chapter 15; Mueller, Chapter 17). Nevertheless, the use of specimens in systematics collections for molecular research has forced the collections community to take a new look at policies and parameters for destructive sampling

(Cato, 1993; Hoagland, 1994). Because researchers often borrow material from several institutions, policies and protocols will be more effective and more likely to be followed if they are consistent among institutions.

The symposium Destructive Sampling and Molecular Systematics: Are We Moving Toward a Consensus? responded to a call for botanists to produce a community recommendation to deal with issues relating to DNA stored in herbaria (Stuessy, 1996), and went a step further by including experiences from other disciplines. In light of the current trend for university and museum biology departments to be clustered around thematic research interests rather than taxon groupings (Roush, 1997; Butler et al., 1998), an understanding of the different disciplinary requirements is fundamental to policy formulation that will serve everyone's needs.

The Harvard University Herbaria were among the first to circulate a specific policy for destructive sampling of herbarium specimens for molecular systematic research. Emily Wood, Torsten Eriksson, and Michael Donoghue (Chapter 15) describe the process that led to the formulation of Harvard's policy. Key to the success of that process was the inclusion of all the constituents — collections staff, morphological systematists, and molecular systematists. Wood et al. highlight a prime herbarium concern that while the value of herbarium specimens is enhanced by their use in molecular systematics, specimen usage cannot compromise the long-term integrity of the specimens. They provide basic guidelines for achieving this balance. Robert Jansen, Dennis Loockerman, and Hyi-Gyung Kim (Chapter 16) detail the pros, cons, and protocols for obtaining DNA from vascular plant specimens. Based on surveys of molecular systematics labs and herbaria, Jansen et al. conclude that herbarium specimens are an underutilized source of material for molecular systematic investigations, but that their use is on the rise. Gregory Mueller (Chapter 17) elaborates on the many ways that fungal collections differ from plant collections, in fact and in usage, creating different sorts of issues for their use in molecular systematics. Based on the results of surveys of herbaria and molecular systematists, Mueller focusses on the implications of specimen preparation and treatment regimes for retrieval of DNA from mycological specimens, and on the pros and cons of herbaria serving as repositories for DNA obtained from their specimens.

James Whitfield (Chapter 18) describes the problems associated with obtaining DNA samples from insect specimens, which, due to their small size, can be completely consumed. He introduces the concept of "remnant voucher material" and calls upon scientific journals to require more stringent documentation criteria for the publication of molecular results. Mark Engstrom, Robert Murphy, and Oliver Haddrath (Chapter 19) note that most vertebrate collections have developed ancillary collections of frozen tissue to service the needs of the molecular systematics community, with the assumption that specimens from these collections will be fully consumed. "Consumptive sampling" of ancillary collections is contrasted with the "destructive sampling" of the main voucher collection. James Miller (Chapter 20) describes analogous ancillary collections of air-dried or desiccant-dried plant material that have been developed at the Missouri Botanical Garden. These collections broaden the base of available material and alleviate problems of DNA amplification that are associated with traditional specimen preservation techniques. Miller empasizes the need for a clear link between the material in the ancillary collection and the herbarium voucher specimen.

The six symposium papers concur that systematics collections are an attractive source of material for molecular systematic research, especially because they are readily accessible and represent a wide array of taxa. They caution that current research needs must be balanced against the long-term integrity of the collections and that clear guidelines are required to ensure that researchers will conform to herbarium/museum etiquette and protocols. The symposium wrap-up session, which involved all presenters and the audience, led to the formulation of a series of recommendations that have been compiled by Deborah Metsger (Chapter 21). They are directed to individual researchers, institutions, and academic societies as guidelines to assist in the formulation of policies and for the creation of standards.

PART III: ABSTRACTS AND SHORT PAPERS
FROM THE HERBARIUM INFORMATION BAZAAR

A fundamental goal of the workshop on which this book is based was to provide a down-to-earth atmosphere with ample opportunity for participants to share information, ideas, and techniques. Part III: Abstracts and Short Papers from the Herbarium Information Bazaar follows from hands-on presentations or posters that were delivered at the workshop. These contributions provide anecdotal accounts of the methods used by various individuals or institutions to address frequently encountered problems associated with specimen preparation, conservation, shipping, storage, and pest control. Many of them complement or illustrate issues raised elsewhere in the book.

CONCLUSION

Williams (1993), in his concluding statement for the 1992 First World Congress on the Preservation and Conservation of Natural History Collections, states: "It was obvious that the most successful initiatives were those that thrived on open communication and cooperation, regardless of differences in nationality, discipline, or professional status. The job of effectively preserving natural history collections is much too massive for any single party to try to do it alone..."

The future of natural history collections of all kinds will depend on interdisciplinary and world-wide collaboration. It is clear that the management and use of collections is no longer the sole concern of individual disciplines. Providing opportunities for the professional development of collections staff is critical to ensuring that they are prepared for the new expanding roles of collections. Cato et al. (1996) note that the field of Collections Management is rapidly evolving to address this need. As expressed by Janes (1995), museums deal in knowledge, and, because knowledge itself grows and changes, there is a responsibility for museum staff to continually adapt and raise the professional standards of their workplace. Professional societies should in turn endorse and disseminate these standards and protocols for specimen usage and encourage their members to practice them (Metsger, Chapter 21).

Sampling the Green World (Stuessy and Sohmer, 1996) speaks to the botanical community at large and puts forth a philosophical challenge to set our goals and priorities for collecting and preserving all kinds of plant materials in line with the trends and realities of the future. *Managing the Modern Herbarium* begins to address how this challenge can be met by taking a practical, problem-solving approach to real situations that are faced by botanists and herbarium staff the world over, on a daily basis. The papers presented in this book reflect the spirit of collaboration and consensus at all levels that we feel is required to move natural history collections in general and herbaria in particular into the future. Enhanced dialogue within the botanical community will lead to standardization of techniques and policies, and reduce duplication of effort. Dialogue between the botanical community and conservators will improve our ability to make informed choices regarding materials, techniques, and building design, while providing a clearer picture of how the specimens are used. This will in turn identify relevant areas for future conservation research. Dialogue between botanists and zoologists will facilitate a better understanding of the discipline-specific requirements for collections and research that must be considered in any collaborative biodiversity initiatives. Dialogue between collections staff and molecular scientists will clarify contemporary research needs and uses, while at the same time explaining the need for protocols and standards that will ensure the long-term integrity of the specimens. Together, these initiatives will help to preserve and promote museum voucher specimens as a vital record, in place and time, of the biological organisms which they represent, and ensure their availability for use by future generations.

ACKNOWLEDGMENTS

We wish to thank K. Barbour, J. Bull, T. A. Dickinson, T. Sage, J. Solomon, J. Waddington, and J. Warren for discussion and/or critical review of this manuscript. This is contribution number 54 from the Centre for Biodiversity and Conservation Biology, Royal Ontario Museum.

LITERATURE CITED

ALBERCH, P. 1993. Museums, Collections and Biodiversity Inventories. TREE 8(10):372-375.

ANDERSON, W. R. 1996. The importance of duplicate specimens in herbaria. Pp. 239-248 in T. F. Stuessy and S. H. Sohmer, editors. *Sampling the Green World*. Columbia University Press, NY, 289 pp.

[AAM] AMERICAN ASSOCIATION OF MUSEUMS. 1998. Challenges and Opportunities. An excerpt from The American Association of Museums' Strategic Agenda. Handout, AAM annual meeting, Los Angeles, CA.

ANONYMOUS. 1994. *Systematics Agenda 2000: Charting the Biosphere*. Technical report. Systematics Agenda 2000, NY.

BLACK, C. C. 1993. Introduction. Pp. 17-20 in C. L. Rose, S. L. Williams, and J. Gisbert, editors. *Current Issues, Initiatives, and Future Directions for the Preservation and Conservation of Natural History Collections. International Symposium and First World Congress on the Preservation and Conservation of Natural History Collections*. Direccion General de Bellas Artes y Archivos, Ministerio de Cultura, Madrid, Spain, vol. 3, 439 pp.

BOOM, B. M. 1996. Societal and scientific information needs from plant collections. Pp. 16-27 in T. F. Stuessy and S. H. Sohmer, editors. *Sampling the Green World*. Columbia University Press, NY, 289 pp.

BRIDSON, D. AND L. FORMAN. 1992. *The Herbarium Handbook*. Revised edition. Royal Botanic Gardens, Kew, 303 pp.

BUTLER, D., H. GEE, AND C. MACILWAIN. 1998. Museum research comes off list of endangered species. Nature 194:115-120.

[CCI] CANADIAN CONSERVATION INSTITUTE. 1994. Framework for Preservation of Museum Collections. Canadian Heritage, Government of Canada, wall chart.

CATO, P. S. 1988. Review of organizations and resources that serve the needs of natural history collections. Collection Forum 4(2):51-64.

CATO, P. S. 1993. Institution-wide policy for sampling. Collection Forum 9(1):27-39.

CATO, P. S., R. R. WALLER, L. SHARP, J. SIMMONS, AND S. L. WILLIAMS. 1996. Developing Staff Resources for Managing Collections. Virginia Museum of Natural History Special Publication Number 4, 71 pp.

CHOLEWA, A. 1997. Problems facing smaller herbaria. Collection Forum 13(1):20-24.

DAVIS, J. I. 1995. Species concepts and phylogenetic analysis – introduction. Syst. Bot. 20(4):555-559.

DUCKWORTH, W. D., H. H. GENOWAYS, AND C. L. ROSE. 1993. *Preserving Natural History Collections: Chronicle of Our Environmental Heritage.* National Institute for the Conservation of Cultural Property, Inc., 140 pp.

FABER, D. J., editor. 1983. *Proceedings of 1981 Workshop on Care and Maintenance of Natural History Collections.* Syllogeus No.44. National Museums of Canada, Ottawa, 196 pp.

GUNN, A. 1994. Past and current practice: the botanist's view. Pp. 11-14 in R. E. Child, editor. *Conservation and the Herbarium.* Institute of Paper Conservation, Leigh, Worcestershire, 41 pp.

HEYWOOD, V. H. 1996. A historical overview of documenting plant diversity: are there lessons for the future? Pp. 3-15 in T. F. Stuessy and S. H. Sohmer, editors. *Sampling the Green World.* Columbia University Press, NY, 289 pp.

HICKS, A. J. AND P. M. HICKS. 1978. A selected bibliography of plant collection and herbarium curation. Taxon 27(1):63-99.

HOAGLAND, E. K., editor. 1994. *Guidelines for Institutional Policies and Planning in Natural History Collections.* Association of Systematics Collections, Washington, DC, 120 pp.

HUMPHREY, P. S. 1991. The nature of university natural history museums. Pp. 5-11 in P. S. Cato and C. Jones, editors. *Natural History Museums: Directions for Growth.* Texas Tech University Press, Lubbock, Texas, 252 pp.

JANES, R. R. 1995. *Museums and the Paradox of Change. A Case Study in Urgent Adaptation.* Glenbow Museum, Calgary, Alberta, 193 pp.

LAWRENCE, G. H. M. 1951. *Taxonomy of Vascular Plants.* MacMillan Publishing Co., NY, 823 pp.

OGILVIE, R. T. 1985. Botanical collections in museums. Pp. 13-22 in E. H. Miller, editor. *Museum Collections: Their Roles and Future in Biological Research.* British Columbia Provincial Museum No. 25, Occasional Papers Series, 222 pp.

ROSE, C. L. 1991. The conservation of natural history collections: addressing preservation concerns and maintaining the integrity of research specimens. Pp. 51-59 in P. S. Cato and C. Jones, editors. *Natural History Museums: Directions for Growth.* Texas Tech University Press, Lubbock, iv + 252 pp.

ROSE, C. L. AND C. A. HAWKS. 1995. The preventive conservation approach to the storage of collections. Pp. 1-20 in C. L. Rose, C. A. Hawks, and H. H. Genoways, editors. *Storage of Natural History Collections: A Preventive Conservation Approach.* Society for the Preservation of Natural History Collections, Iowa City, IA, 448 pp.

ROSE, C. L. AND A. R. DE TORRES, editors. 1992. *Storage of Natural History Collections: Ideas and Practical Solutions.* Society for the Preservation of Natural History Collections, Iowa City, IA, 346 pp.

ROSE, C. L., C. A. HAWKS, AND H. H. GENOWAYS, editors. 1995. *Storage of Natural History Collections: A Preventive Conservation Approach.* Society for the Preservation of Natural History Collections, Iowa City, IA, 448 pp.

ROSE, C. L., S. L. WILLIAMS, AND J. GISBERT, editors. 1993. *Current Issues, Initiatives, and Future Directions for the Preservation and Conservation of Natural History Collections.* International Symposium and First World Congress on the Preservation and Conservation of Natural History Collections. Direccion General de Bellas Artes y Archivos, Ministerio de Cultura, Madrid, Spain, vol. 3, 439 pp.

ROUSH, W. 1997. Biology departments restructure. Science 275:1556-1558.

SAVILE, D. B. O. 1962. Collection and care of botanical specimens. Canadian Dept. of Agriculture Publication 1113, Ottawa, 124 pp.

STEARN, W. T. 1957. An introduction to the Species Plantarum and cognate botanical works of Carl Linnaeus – in vol. 1, Carl Linnaeus – *Species Plantarum* – A facsimile of the first edition 1753. The Ray Society, London.

STERN, M. J. AND T. ERIKSSON. 1996. Symbioses in herbaria: recommendations for more positive interactions between plant systematists and ecologists. Taxon 45:49-58.

STUESSY, T. F. 1996. Are present sampling, preservation, and storage of plant materials adequate for the next century and beyond? Pp. 271-284 in T. F. Stuessy and S. H. Sohmer, editors. 1996. *Sampling the Green World.* Columbia University Press, NY, 289 pp.

STUESSY, T. F. AND S. H. SOHMER, editors. 1996. *Sampling the Green World.* Columbia University Press, NY, 289 pp.

WADDINGTON, J. AND D. M. RUDKIN. 1986. *Proceedings of the 1985 Workshop on Care and Maintenance of Natural History Collections.* Royal Ontario Museum Life Sciences Miscellaneous Publications, Toronto, 121 pp.

WILLIAMS, S. L. 1993. Closing Remarks of the First World Congress. Pp. 433-435 in C. L. Rose, S. L. Williams, and J. Gisbert, editors. *Current Issues, Initiatives, and Future Directions for the Preservation and Conservation of Natural History Collections.* International Symposium and First World Congress on the Preservation and Conservation of Natural History Collections. Direccion General de Bellas Artes y Archivos, Ministeriio de Cultura, Madrid, Spain, vol. 3, 439 pp.

WILLIAMS, S. L. AND P. S. CATO. 1995. Interaction of research, management, and conservation for serving the long-term interests of natural history collections. Collection Forum 11(1):16-27.

PART I

PREVENTIVE CONSERVATION

AND

COLLECTIONS MANAGEMENT

SECTION 1

PREVENTIVE CONSERVATION:

PRINCIPLES AND PRACTICE

CHAPTER 1

Preventive Conservation: Concept and Strategy for Long-Term Preservation

JOHN TOWNSEND

Abstract.—Preventive conservation is a management tool designed to help prolong the useful life of collections. It attempts to identify the likely causes of deterioration of collections, discern the interactions that occur among the causal agents and grasp the limitations imposed on the institution by its location, facilities, and resource capabilities. An integrated approach to preventive conservation management will facilitate an understanding of these issues. By following this approach an institution can systematically address its responsibility, need, decision, and means to preserve its collections. A conservation assessment, if properly conducted, will place all aspects of a collections history, condition, care, use, and vulnerability within an institutional, environmental, and cultural context. The resulting information will guard against both the dangerous application of packaged solutions to conservation issues and the continuance of harmful practices based on past traditions. The final product will be a set of tools for developing effective long-term strategies for collection care that are customized to meet the specific needs of a given herbarium.

INTRODUCTION

Herbaria, perhaps even more so than other natural history collections, often follow collections management practices that have been handed down from generation to generation. These practices have usually been formulated on the basis of convenience, availability of materials, and either discipline convention, or word of mouth, rather than the recommendations of conservators or conservation literature. It is thus not surprising that there are few specific guidelines or rules for the conservation of herbarium collections. Neither is there a single solution to herbarium conservation problems. Rather, the development of a conservation management scheme for any given herbarium must draw upon the resources and experience of many different disciplines. Given the nature of herbarium specimens—organic material, sometimes with viable parts, mounted with accompanying labels in or on paper products with adhesive—conservation measures must take into consideration all of these materials and their complex interactions. A management scheme using an integrated approach to the assessment of actual and potential problems associated with herbarium collections, within the context of the institutional mandate, will facilitate the determination of priorities for their appropriate care and maintenance. Before elaborating further on the circumstances of specific herbaria, let us first review the principles of preventive conservation.

PREVENTIVE CONSERVATION: THE CONCEPTUAL BASIS

Preventive conservation, at the conceptual level, attempts to manage the inevitable deterioration of collections through time by slowing its progress. Its goal is to minimize the impact on the nature of the collections, and correspondingly, the mandate of the institution. While the principles of preventive conservation are quite straightforward, their practical application is more complex. It requires institutions to assess their priorities and resources, and to find active solutions to what may be large-scale and serious threats to the integrity of their collections. Having done this, they must then determine the most likely agents of deterioration for a particular collection and the interplay between these agents. Further, the institution must identify the constraints to action imposed by its geographic location, its physical structures and, not least, its financial capability to implement and sustain long-term preservation. Unfortunately, institutional plans for preventive conservation often continue to ignore these important local constraints in favor of inflexible, dogmatic, and ultimately uninformed approaches to preventive conservation. This approach is sustained by the myth that preservation problems are universal and that pre-packaged solutions are available. In reality, plans for preventive conservation must be managed and integrated in the same manner as any other operation within an institution.

An integrated approach to managing collections differs from what could be considered a traditional approach of deferred maintenance, that is, reacting to a problem once it occurs rather than trying to prevent its occurrence by understanding the causal factors. An integrated approach also places preventive conservation within the context of the particular institution, and correspondingly the institution within the prevailing social, cultural, and economic context. The process of achieving such an approach is multi-tiered. It must begin with a detailed literature review of analogous situations and effective preventive conservation programs, with an eye to extracting applicable strategies. The appropriate approach for a given collection must include an honest assessment of the following questions:

- Is there a recognized institutional *responsibility* for these collections?
- Is there a *need* to conserve these collections?
- Is there an institutional *decision* and commitment to conserve all or part of these collections?
- Does the institution have the *means* to conserve these collections?

The way an institution responds to these inter-related questions reflects both its ability and its willingness to incorporate a preventive conservation program as a priority within its overall operations.

The *responsibility* for the long-term preservation of collections

ROSEWARNE LEARNING CENTRE

is a prerequisite to developing a preventive conservation program. This responsibility is determined by factors that define the institutional mandate, e.g., its mission, goals, collecting mandates, cultural and historical context, and any cooperative agreements with sister institutions.

Given its responsibility to these collections, the decision to conserve them is predicated on the *need* to conserve. The determination of need includes an evaluation of collections use, and a critical assessment of the current state of preservation of the collections and the potential for future deterioration.

Once the need has been determined, an institution's administration must make a *decision* to conserve its collections prior to the development of a preventive conservation plan. Such a plan requires sufficient funds to support it and therefore a certain commitment to do so. Thus, an institution's commitment to support a preventive conservation plan requires the overall *means* to do so. Resources including funding, staffing, facilities, and appropriate technologies are *all* paramount.

The availability of these resources cannot be taken for granted especially in the prevailing climate of budget constraints for scientific and cultural organizations, where the administrative allocation of resources for preventive conservation may be accompanied by corresponding liabilities such as a reduction in funding for collecting, sponsored research, or staffing. If, as sometimes happens, preservation is perceived as not having the high profile needed to attract additional support, preventive conservation may not receive the institutional support it needs. Ironically, the failure to adequately fund preventive conservation measures will almost certainly perpetuate the problem of failing to notice preservation problems until collections are already in jeopardy. At this point the problems are usually more difficult (and more expensive) to solve.

THE CONSERVATION ASSESSMENT

The process of answering the questions posed above is often termed a preservation (conservation) assessment (Wolf, 1990). Whether conducted by in-house staff or by an outside consultant, such an assessment offers institutions of all sizes "a process, or set of tools for developing long-term strategies for collections care" (Wolf, 1993).

While there is, to my knowledge, no assessment format specific to herbaria, there are a number of formats for other types of institutions that may be adapted and used (National Park Service, 1990; Wolf, 1990; Michalski, 1992; Reed-Scott, 1993). They all work in more or less the same way, providing a structure for identifying and evaluating the specific conservation needs of an institution's collections. The ultimate goal is to provide a mechanism to enable the institution's staff "to establish conservation as an integral part of the [its] mission..." (Wolf, 1993). An assessment includes all of the components of the conceptual framework for preventive conservation: namely, an evaluation of building(s) and facilities; climate control systems and environmental conditions; collections and collection policies; storage facilities; risks to the collections and associated facilities; emergency preparedness; and, resources available for meeting defined preservation needs. The assessment thus provides the foundation for the practical resolution of conservation issues.

AGENTS OF DETERIORATION IN COLLECTIONS: ACTIONS AND INTERACTIONS

The mechanisms of deterioration of collections and the procedures for dealing with them are described in detail in standard works on the curation and preservation of collections (e.g., Thomson, 1986; Michalski, 1992; Thompson, 1992). These references should be used to discern possible solutions to a given problem rather than adopted wholesale. Reiterating the caution against packaged solutions already discussed, the appropriate solution may be an amalgamation of many elements combined to meet the circumstances of the collection in question.

The agents that cause collections to deteriorate can be grouped into one of four categories: biological (e.g., insects, rodents, mold); environmental (e.g., temperature, relative humidity, light and air quality); chemical (e.g., acid hydrolysis); and physical (e.g., mechanical damage from handling of specimens). Any single factor is sufficient to cause some degree of deterioration, and, in extreme circumstances, irreparable damage or complete loss of collection materials. However, they rarely act alone. There is a matrix of interlocking agents acting together, rather than four discrete independently acting agents. Some interactions are not easily anticipated, and so may be a result of our attempts to manage a particular agent of deterioration without regard to the implication of our actions on the effect(s) of other agents.

To further complicate the picture, the impact of any combination of agents on the deterioration of collections will vary with geographic location and cultural context. For example, high relative humidity (RH) is seldom a problem in the southwestern United States whereas it is a significant seasonal problem in the southeastern United States. The situation, however, may be reversed within buildings where microclimates are established in the attempt to achieve adequate levels of heating or cooling. Consequently, in the southwest, high RH may result from cooling buildings, while in the southeast RH levels may be damagingly low in the winter due to inadequate humidification while heating. Sometimes opposite extremes of RH can even be found within the same building due to structural or floor plan modifications incompatible with the heating, ventilation, and air conditioning (HVAC) system.

How do the complex interactions amongst agents of deterioration affect the development of a preventive conservation management plan, and what are the consequences of failing to recognize them? Two examples based on recent experience at the Herbarium Bogoriense (BOGOR), a unit of the Research and Development Center for Biology in Bogor, Indonesia can be used to illustrate. Both examples suggest that there are unexpected ways in which agents of deterioration can interact, and, just as importantly, how attempts to solve perceived problems may further complicate efforts to manage the deterioration of collections.

The first example deals with paper deterioration and insects, an interplay between chemical and biological agents. As background, the potential for the deterioration of paper in an herbarium is a critical concern because mounting sheets, labels, and other paper products are an integral, permanent part of the research and collection record. Unlike most other natural history collections, herbaria do not maintain a separate catalogue. Thus prior to computerization, the only record of data for a specimen was on the mounting sheet itself. Any deterioration that compromises the integrity of this record is therefore unacceptable. To reduce the risk of paper deterioration, it has in recent years become standard practice for most herbaria to strive to use only archival-quality paper for mounting all new specimens or remounts. Ideally, paper materials should meet the minimum standards of permanence (i.e., the ability to resist chemical degradation) and durability (i.e., the ability to retain its physical properties when subjected to stress) as established by the American National Standards Institute (1992) or similar standards established by other organizations.

Unfortunately, improvements to paper quality have no impact on the threat to specimen integrity posed by insects. Insects that eat paper are equally satisfied with good or poor quality paper. Furthermore, the more serious threat is from insects that eat the actual plant material (e.g., *Lasioderma*) or adhesives. The only effective means of deterring the destructive capabilities of these organisms is to seal the mounted specimens within a metal case. Even high quality cases, however, are an inadequate barrier to insect infestation due to the regular examination, insertion, or removal of specimens. Consequently, a program of preventive conservation is still required to deal with the biological problem of potential insect infestation, even after the chemical and physical threats from paper deterioration have been addressed.

At BOGOR numerous mounting sheets and other papers were observed to have turned a deep purplish-brown and collapsed under the weight of the specimen. The degree of degradation far exceeded anything normally observed as a result of typical paper acidity or poor storage conditions. The fragility of some mounting sheets was comparable to paper that had been charred by fire. Some of the most dramatic examples were of fairly recent date (within the last 30 years). In contrast, other examples of the same mounting papers (identified by watermark) were somewhat fragile but, even when twice as old, they still supported their specimens and lacked the characteristic discoloration.

What was the cause of such rampant deterioration? A review of specimen treatment indicated that mercuric chloride was used in the collections at BOGOR until late 1994. Treatment included application of this insecticide to the plants prior to mounting and, in some cases, application to the plant and/or the sheet after mounting. Mercuric chloride was thought to be an excellent choice as an insecticide for herbaria because it is lethal to all the insects that attack plant collections, and it has residual effects that protect the specimens against future infestations. The literature suggests that this practice was particularly common in tropical herbaria, although some authors caution against application of the chemical directly to the mounting sheet (Eusebio and Stearn, 1964; Fosberg and Sachet, 1965; Womersley, 1981). This caution is based on the understanding that mercuric chloride interacts with the acid-producing compounds used in the manufacture of paper (Fosberg and Sachet, 1965; Womersley, 1981; Clark, 1988; Moore, 1990) and results in accelerated degradation. This would suggest that the cause of the extreme cases of deterioration and discoloration of the mounting sheets at BOGOR was a result of the application of mercuric chloride, especially that applied to the sheet subsequent to mounting.

However, it is not just this additional (and in some cases re-peated) application of mercuric chloride that is the problem. In addition to paper deterioration, some plant specimens that had been treated with mercuric chloride were also quite brittle, result-ing in excessive fragmentation during remounting, in spite of very careful handling. These exceptionally brittle plant specimens were observed in some instances when their mounting sheets did not appear to have received direct application of mercuric chloride, that is, the paper showed no more than normal deterioration. This suggests that even routine use of mercuric chloride to treat plant material alone is inappropriate for research collections.

These observations are not new or unique. They confirm what we presently know about the effects of mercuric chloride: it is in-compatible with the goals of long term preservation of plant speci-mens, including their mounts and labels, even though it may be effective in protecting them against attack by insects. It would seem, therefore, a simple matter to avoid the undesirable conse-quences of treatment with mercuric chloride by employing one of the more satisfactory alternatives for pest control. However, the connection between one problem and another is not always perceived with sufficient clarity to influence this type of decision-making.

This lack of clarity was illustrated at BOGOR when a team of visiting researchers vigorously protested the cessation of the treatment of herbarium specimens with mercuric chloride for the sake of human health and safety. Their heated arguments were based on two firmly held assumptions: first, that the value of the collections was so great that human health and safety, including their own, should not be allowed to compromise the safety of the collections; and second, that mercuric chloride is the absolute best choice to preserve tropical collections from the ravages of insect pests. Though admittedly extreme, this is a perfect example of how traditional approaches to specimen preservation are sometimes treated as sacrosanct and are not subject to the rational and scientific scrutiny routinely applied to other areas of inquiry. Leaving aside the social, moral, and ethical implications of these assumptions, it is clear that they are flawed on scientific grounds. It also illustrates how the failure to account for the obvious interaction among different agents of deterioration can work against even the most zealous attempts at preservation.

The second example is less disturbing but more complex. It has to do with the interactions between environmental, biological, and physical agents, namely, temperature, RH, mold, and building design. The recommended set points for HVAC systems are still debated among conservation professionals (Thomson, 1986; Michalski, 1993; Sebera, 1994). There is, however, wide agreement on both the acceptable ranges for most types of collection materials, and on the adverse effects of cycling, or quick fluctuations in conditions (Thomson, 1986; Sebera, 1989, 1994; Reilly et al., 1995). Due to the relative lack of research into, and scientific certainty about the complex interactions amongst environmental and other agents of deterioration, it is difficult to evaluate the environmental conditions for a given location, let alone prescribe ready-made solutions.

In temperate climates, RH levels at or above 70% are regularly observed to result in outbreaks of mold growth on collections in libraries, archives, museums, and other institutions. By contrast, large-scale outbreaks of mold on collections appear to be uncommon in tropical climates like that of Bogor and other Indonesian cities where RH is nearly always above 70%. Recent personal observations at institutions in Bogor, Bandung, Jakarta, Yogyakarta, Solo, Denpasar, and Singaraja found active mold growth on collection materials only when they had been directly damaged by water leaks, or were held in newer, western-style buildings with modern ventilation systems. Based on temperate experience this seems illogical. Additional agents besides RH must be at work to prevent what would otherwise be expected as a major preservation problem in the older buildings, yet cause it in the new buildings.

The most probable explanation lies in building design. Many of the institutional buildings cited are of earlier construction where ventilation was provided through open windows. These buildings typically have thick masonry walls and high ceilings and roofs with wide overhangs that limit direct sunlight and reduce heat gain. Modern buildings observed in Indonesia, by contrast, do not always have these features and frequently rely on intermittent air conditioning for cooling and ventilation. As in many other developing countries the high cost of electrical generation often precludes 24-hour operation of the ventilation system for even the most well-funded institutions. When the air conditioning in these buildings is not in use, the temperature may rise sharply during the course of a day, as a function of heat gain either through large windows that do not open, or thinner walls that are not shaded. When the air conditioning is re-started the temperature drops rapidly. Due to the natural inter-dependence of temperature and RH this dramatic fluctuation is likely the cause of condensation and other moisture problems (including mold) typical of these modern western-style buildings. If these observations are accurate they suggest that in tropical countries, unless adequately designed HVAC systems are operated on a 24-hour basis, modern building designs reliant on mechanical systems for climate control are less appropriate than those of indigenous and earlier colonial design that provide better air circulation.

What about other aspects of the tropical condition? Based on research and experience in temperate climates one would normally expect Bogor's tropical climate (which includes nearly constant rainfall, in excess of 4,300 mm per annum) to be significantly more damaging overall to paper and other organic materials. As mentioned above, this does not appear to be the case. Why? Long-term meteorological data indicate that Bogor itself has a very stable climate with relatively constant temperatures and RH throughout the year. This is in sharp contrast to the extreme seasonal and yearly fluctuations characteristic of many temperate climates. It is also in contrast to the conditions described in modern Indonesian buildings. Perhaps the natural stability of Bogor's climate has a greater significance than the magnitude of any individual environmental element. If so, then one could postulate that the stability of the environment has protected the Herbarium collections from the level of deterioration that research and experience in temperate environments would predict. Thus, the installation of limited environmental controls in an old-style herbarium building would likely serve to disrupt environmental conditions rather than stabilize them. The result could well lead to the destruction rather than preservation of specimens. While not recommending air conditioning may seem heretical from a western perspective, even casual attention to prevailing local conditions in Bogor makes it painfully obvious that the application of a prescriptive, packaged recommendation based on temperate conditions is completely inappropriate.

CONCLUSION

Both of the examples given emphasize the importance of keeping preventive conservation efforts in perspective with their local institutional, cultural, and environmental context. It is detrimental to try to combat interacting agents that are not causing problems or, conversely, to ignore the interaction between agents while treating only the most obvious one. Preventive conservation is a process. It is a balancing act, juggling the priorities, needs, and capabilities of any given institution. If it is to be effective it must be practical, and if it is to be practical then it must take into account conditions as we find them, where we find them, and address them with the resources at hand. The information we need to address the specifics of preventive conservation is available in the professional literature, in our careful observations of conditions at a particular location, and in the accumulated wisdom and experience of researchers, technicians, curators, managers, and others. To forge this diverse information into an effective conservation management tool requires a fresh perspective, one that recognizes that there are few, if any, absolutes in preservation but rather tremendous opportunities to use informed creativity and common sense to devise workable solutions for extending the life of collections.

Literature Cited

American National Standards Institute. 1992. *Permanence of Paper for Publications and Documents in Libraries and Archives: ANSI Standard Z39.48.*

Clark, S. 1988. Preservation of herbarium specimens: an archival approach. Library Conservation News 19:4-6.

Eusebio, M. A. and Stearn, W. T. 1964. Preservation of herbarium specimens in the humid tropics. Philippine Agriculturist 48(1):16-20.

Fosberg, F. R. and Sachet, M.-H. 1965. Manual for Tropical Herbaria. Regnum Vegetabile 39:1-132.

Michalski, S. 1992. *A Systematic Approach to the Conservation (Care) of Museum Collections.* Canadian Conservation Institute, Ottawa.

Michalski, S. 1993. Relative humidity: A discussion of correct/incorrect values. Pp. 624-629 in *Preprints of ICOM Conservation Committee 10th Triennial Meeting*, Washington, DC, 22-27 August 1993: International Council of Museums Committee for Conservation, Paris.

Moore, B. P. 1990. *Care and Conservation of the Collections of the Peabody Museum of Natural History.* Unpublished report, Peabody Museum, Yale University.

National Park Service. 1990. *Museum Handbook Part I: Museum Collections.* National Park Service, Washington, DC.

Reilly, J. D., D. Nishimura and E. Zinn. 1995. *New Tools for Preservation: Assessing Long-Term Environmental Effects on Library and Archives Collections.* The Commission on Preservation and Access, Washington, DC.

Reed-Scott, E. 1993. *Preservation Planning Program: An Assisted Self-Study Manual for Libraries.* Second edition. Association of Research Libraries: Washington, DC.

Sebera, D. K. 1989. A graphical representation of the relationship of environmental conditions to the permanence of hygroscopic materials and composites. Pp. 51-75 in *Proceedings of Conservation in Archives: International Symposium*, Ottawa, Canada, May 10-12, 1988: International Council on Archives, Paris.

SEBERA, D. K. 1994. *Isoperms: An Environmental Management Tool.* Commission on Preservation and Access, Washington, DC.

THOMPSON, J. M. A., editor. 1992. *Manual of Curatorship: A Guide to Museum Practice.* Second edition. Butterworths, London.

THOMSON, G. 1986. *The Museum Environment.* Second edition. Butterworths, London.

WOLF, S. J., editor. 1990. *The Conservation Assessment: A Tool for Planning, Implementing, and Fund Raising.* Getty Conservation Institute and the National Institute for the Conservation of Cultural Property, Washington, DC.

WOLF, S. J. 1993. Conservation assessments and long-range planning. Pp. 289-307 in C. L. Rose, S. L. Williams, and J. Gisbert, editors. *Current Issues, Initiatives, and Future Directions for the Preservation and Conservation of Natural History Collections.* International Symposium and First World Congress on the Preservation and Conservation of Natural History Collections. Dirección General de Bellas Artes y Archivos, Ministerio de Cultura, Madrid, Spain, vol. 3, 439 pp.

WOMERSLEY, J. S. 1981. *Plant Collection and Herbarium Development.* United Nations Food and Agriculture Organization, Rome.

CHAPTER 2

Conservation and Collections Care Resources

CAROLYN L. ROSE

Abstract.—Research, management, and conservation must work in concert to best serve the long-term interests of natural history collections and to insure their continued viability. A good understanding of current conservation principles, practices, and approaches provides a means to achieving this goal. The use of available knowledge and technology offered through publications and educational forums can avoid costly mistakes and the duplication of efforts, and will assist in preserving valuable herbarium collections for the future. Some of the available conservation and collections care resources are summarized, by category, with special attention to those that could be useful to herbarium personnel.

INTRODUCTION

Important interrelationships must exist among research, management, and conservation in order to serve the long-term interests of natural history collections. Research gives purpose to the collections, and justification for management and conservation; management provides the mechanisms to fulfill that purpose; and conservation provides a responsible approach to preserving the collections and to maintaining the research integrity of specimens. All three areas must work in concert, providing a balance among use, management, and conservation activities (Williams and Cato, 1995). For those who are responsible for balancing these activities for herbarium collections, this chapter provides information on useful conservation and collections care resources, especially those in North America.

Conservation involves four basic activities: preventive care, treatment, documentation, and research. Traditionally, conservation activities focused on the examination, treatment, and analysis of individual works of art. In recent years, however, the emphasis has shifted to preventive care involving risk assessment and the development of strategies to mitigate deterioration. With this shift in approach have come new efforts to involve other museum personnel in the collections care process particularly in the underrepresented area of natural history collections. As a result, there has been a rapid growth of publications, workshops, and other resources to enable all collecting institutions, arts and science alike, to provide professional care for their collections. Many of these resources are provided by conservation/preservation organizations and institutions, or by collection-oriented associations that are concerned with the long-term preservation of collections as part of broader collection or research goals. Useful information is also available from standards and testing organizations, health and safety organizations, and from suppliers and manufacturers of conservation equipment and materials. For example, several of the conservation product catalogues contain useful descriptions of the composition of storage and mounting materials, guidelines for selecting appropriate materials, and glossaries of conservation terms.

Before selecting specific products for herbarium collections, however, it is wise to verify product claims with a conservator and/or conservation scientist. These professionals can be contacted through one of the conservation professional organizations or through a conservation or preservation research laboratory. The Conservation Committee and publications of the Society for the Preservation of Natural History Collections (SPNHC) can also provide valuable information on the usefulness of suggested products for herbarium collections (Rose and Torres, 1992; Rose, 1993; Rose et al., 1995).

SELECTED CONSERVATION/
PRESERVATION RESOURCES

The names and addresses of organizations that provide conservation resources are listed below in topical sections. Each section begins with an overview of the categories of resources included and their general application to collections. To assist the reader in assessing which resources are relevant to their needs, entries include lists of publications, education and training opportunities, services, and relevant committees. Please note that addresses of organizations may change every three or more years when new coordinators are elected.

International and National Conservation
and Preservation Professional Associations

The International Institute for the Conservation of Historic and Artistic Works (IIC), founded in 1950, was the first international professional organization for conservators. After the IIC was established, a number of regional groups were founded in various countries. In subsequent years, some of these regional groups became separate professional organizations.

These professional associations of conservators, conservation scientists, and educators provide important foci for collecting and disseminating conservation information. Some associations have a code of ethics and standards of practice which guide the members. Special membership status such as "Fellow" can signify that the member adheres to these standards and has achieved a high level of competency through education, training, and experience. Other organizations may provide certification for conservators. When choosing a conservator to assist in developing a collections care program or to carry out conservation treatments, it is important to know what status the conservator holds in a professional organization, and what that status means (usually explained in societal membership information).

The Society for the Preservation of Natural History Collections, founded in 1985, has also been included in this category because it is a professional organization that provides a bridge between conservation/preservation interests and curation.

American Institute for Conservation of Historic and Artistic Works (AIC)

1717 K St. NW, Suite 301, Washington, DC 20006, USA; (202)452-9545; Fax: (202)452-9328
• PUBLICATIONS: newsletter *(AIC Newsletter)*, journal *(Journal of the American Institute for Conservation)*, Code of Ethics, Standards of Practice, annual membership directory, information brochures, abstracts of papers presented at annual meetings, publications compiled by specialty groups
• EDUCATION/TRAINING: training courses, workshops, refresher courses, annual meetings
• SPECIALTY GROUPS: workshops and publications provided by the following groups: Architecture, Book and Paper, Objects, Paintings, Photographic Materials, Textiles, Research and Technical Studies, Wooden Artifacts
• SERVICES: Conservation Services Referral System (brochure available)

Asociación para la Conservación del Patrimonio Cultural de las Americas (APOYO)

c/o Amparo Rueda de Torres, APOYO, PO Box 76932, Washington, DC 20013, USA; ator@loc.gov
• PUBLICATIONS: newsletter *(APOYO)*

Canadian Association For Conservation of Cultural Property (CACCP, formerly IIC–CG, International Institute for Conservation of Historic and Artistic Works)

PO Box 9195, Ottawa, ON K1G 3T9, Canada; (613)998-3721; Fax: (613) 998-4721

• PUBLICATIONS: newsletter *(CAC Bulletin)*, journal *(Journal of CAC)*, Code of Ethics, annual reports
• EDUCATION/TRAINING: informal conservation training
• OTHER: regional coordinators located across the country

Canadian Association of Professional Conservators (CAPC)

#400, 280 Metcalfe St., Ottawa, ON K2P 1R7, Canada; (613)998-4971; Fax: (613)993-3412
• PUBLICATIONS: newsletter/membership *(Directory)*, informational brochures, Code of Ethics (in conjunction with the former IIC–CG, now CACCP)
• SERVICES: professional certification for conservators, referral service

Institute for Paper Conservation (IPC)

Leigh Lodge, Leigh, Worcestershire WR6 5LB, UK; 44(01886)832323; Fax: 44(01886)833688
• PUBLICATIONS: newsletter *(Paper Conservation News)*, journal *(The Paper Conservator)*, books (e.g. *Conservation and the Herbarium*, 1994)
• EDUCATION/TRAINING: annual meetings, international conferences
• SERVICES: conservation advice

International Council of Museums (ICOM): Committee for Conservation (ICOM–CC)

Maison de l'Unesco, 1 rue Miollis, 75732 Paris XVe, France; 33(1)734-05-00; Fax: 33(1)43-06-78-62
• PUBLICATIONS: newsletters of working groups, preprints, Code of Ethics, books
• EDUCATION/TRAINING: training in

conservation and restoration,
workshops, conferences
• WORKING GROUPS: newsletters,
triennial meeting preprints,
miscellaneous publications

Relevant working groups include:
• Ethnographic Collections
Sherry Doyal, Co-coordinator, Royal
Albert Memorial Museum, Exeter, UK;
Fax: 44(1) 392-421252;
Nancy Odegaard, Co-coordinator,
Arizona State Museum, Tuscon, AZ
85721, USA; Fax: (520)621-2976
• Natural History Collections
Ann Pinzl, Nevada State Museum,
Carson City, NV 89710-0001, USA;
Fax: (702)687-4168
• Photographic Records
Bertrand Lavedrine, Centre de
Recherches sur la Conservation des
Documents Graphiques, Musée
National d'Histoire Naturelle,
36 rue Geoffroy-Saint-Hilaire, F-75005
Paris, France; 33(1)45-87-06-12;
Fax: 33(1)47-07-62-95
• Preventive Conservation
Stefan Michalski, Coordinator,
Canadian Conservation Institute,
Ottawa, ON K1A 0C8, Canada;
(613)998-3721; Fax: (613)998-4721
• Training in Conservation and Restoration
Kathleen Dardes, Coordinator, Getty
Conservation Institute, 1200 Getty
Center Drive, Los Angeles, CA 90049-
1684, USA; Fax:(310)440-7702

International Institute for Conservation of Historic and Artistic Works (IIC)
6 Buckingham St., London WC2N
6BA, UK; 44(171)839-5975;
Fax: 44(171)976-1564
• PUBLICATIONS: newsletter *(Bulletin)*,

journal *(Studies in Conservation)*, books
• EDUCATION/TRAINING: conferences,
courses, seminars

Regional Groups:
• Grupo Espanol Del IIC
% IPHE, C/Greco 4, 28040 Madrid,
Spain
• IIC Hellenic Group
PO Box 27031, 117 02 Athens, Greece
• IIC Japan
Tokyo National Research Institute of
Cultural Properties, 13-27 Ueno Park,
Taito-ku, Tokyo 110, Japan
• IIC Netherland
Robien van Gulik, Ruys de
Beerenbroucklaan 54, 1181 XT
Amstelveen, Netherlands
• Nordisk Konservatorforbund—IIC
Nordic Group
Kari Greeve, Nasjonalgalleriet,
Postboks 8157–Dep, 0033 Oslo,
Norway
• Osterreichische Sektion des IIC
Restaurier—Werkstätten, Arsenal
Objekt 15/Tor 4, 1030 Wien, Austria
• Section Française de l'IIC
29 rue de Paris, 77420 Champs sur
Marne, France

Natural Sciences Conservation Group
Kate Andrew, c/o Ludlow Museum,
Old St., Ludlow, Shropshire SY8 1NW,
UK
• PUBLICATIONS: newsletter *(Natural Sciences Conservation Group)*
• EDUCATION/TRAINING: workshops

Society for the Preservation of Natural History Collections (SPNHC)
Lisa Palmer, Treasurer, PO Box 797,
Washington, DC 20044-0797, USA;

(202) 786-2426; Fax: (202) 357-2986;
palmer.lisa@nmnh.si.edu
• PUBLICATIONS: newsletter *(SPNHC
Newsletter)*, journal *(Collection Forum)*,
professional guidelines, technical
information (e.g. leaflets), books (e.g.
*Storage of Natural History Collections,
Vol. I: A Preventive Conservation
Approach,* 1995 and *Storage of Natural
History Collections, Vol. II: Ideas and
Practical Solutions,* 1992, reprint 1995)
• EDUCATION/TRAINING: annual
conferences; interdisciplinary and
discipline-specific workshops, symposia
on collections management, care and
conservation
• SERVICES: liaison between the
conservation and preservation
community and the various natural
science disciplines

Relevant Committees:
• Conservation Committee
Diane Dicus, Co-Chair, # 205, 1445
Camel Back Lane, Boise, ID 83702,
USA; (208)331-0287;

Fax: (208)384-8567
• Documentation Committee
Jackie Zak, Chair, Getty Conservation
Institute, 1200 Getty Center Drive,
Suite 700, Los Angeles, CA 90049,
USA; (310)440-6226;
Fax: (310)440-7710; jzak@getty.edu
• Education and Training Committee
Iris Hardy, Chair, Geological Survey of
Canada, Bedford Institute of
Oceanography, 1 Challenger Blvd., PO
Box 1006, Dartmouth, NS B2Y 4A2,
Canada; (902)426-6127; Fax:
(902)426-6186; Hardy@agc.bio.ns.ca

United Kingdom Institute for Conservation (UKIC)

6 Whitehorse Mews, Westminster
Bridge Rd., London SE1 7QD, UK;
44(171)620-3371; Fax: 44(171)620-
3761
• PUBLICATIONS: newsletter
(Conservation News Grapevine), journals
(The Conservator), course booklets
(including conservation courses and
training available in the UK), papers
published from conferences

US Regional Conservation Professional Associations

Conservators throughout the United States have formed regional associations and guilds to serve as a forum for information
exchange with local colleagues. Because these regional groups are
relatively small and members are frequently in close communication, the president or secretary of the association can usually make
referrals for specific questions about the preservation of documents and herbarium collections (see also AIC, 1997).

Chicago Area Conservation Group

2600 Kelsinger Rd., Geneva, IL 60134,
USA; (630)232-1708; Fax: (630)208-
1207; Craig1708@aol.com

Conservation Associates of the Pacific Northwest

PO Box 2756, Olympia, WA 98507-
2756, USA; Fax: (360)754-2093;
stilsonr@elwha.evergreen.edu

**Louisiana Art Conservation
Alliance**
PO Box 71473, New Orleans, LA
70172-1473, USA

**Midwest Regional Conservation
Guild**
Indiana University Art Museum,
Conservation Dept., Bloomington, IN
47405, USA; (812)855-1024; Fax:
(812)855-1023; dthimme@indiana.edu

**New England Conservation
Association**
24 Emery St., Medford, MA 02155,
USA; (617)396-9495

**New York Conservation
Association**
345 8th Ave., # 15A, New York, NY
10011, USA

**Upper Midwest Conservation
Association**
2400 3rd Ave. South, Minneapolis,
MN 55404, USA; (612)870-3120;
Fax: (612)870-3118

Virginia Conservation Association
PO Box 4314, Richmond, VA 23220,
USA; (804)358-7545

Washington Conservation Guild
PO Box 23364, Washington, DC
20026, USA; (301)238-3700 ext. 178

**Western Association for Art
Conservation**
c/o Chris Stavroudis, 1272 North
Flores St., Los Angeles, CA 90069,
USA; (213) 654-8748; Fax:(213)656-
3220; estavrou@netcom.com

**Western New York Conservation
Guild**
51 Park Lane, Rochester, NY 14625,
USA; (716)248-5307

US Regional Conservation/Preservation Centers

In the United States, a number of regional conservation centers have been developed to provide conservation services such as assessments, surveys, and treatments to institutions and private collectors. Several centers employ paper conservators and/or ethnographic conservators who can assist herbarium personnel. A number of these centers are partially funded by institutional membership. Subscribing museums and institutions may receive regular information and either free or discounted services.

The Amigos Preservation Service
Amigos Bibliographic Council, 12200
Park Central Dr., Suite 500, Dallas, TX
75251, USA; (800)843-8482;
(214)851-8000; Fax: (214)991-6061;
amigos@amigos.org

• PUBLICATIONS: informational handouts
and articles, bibliographies, books,
videotapes
• EDUCATION/TRAINING: workshops
• SERVICES: regional library and archival
consulting, training, disaster assistance

Balboa Art Conservation Center

PO Box 3755, San Diego, CA 92163,
USA; (619)236-9702;
Fax: (619)236-0141
• PUBLICATIONS: newsletters *(BACC Newsletter),* leaflets (technical
information, orientation to BACC
services and structures)
• EDUCATION/TRAINING: pregraduate
(volunteers and assistants), graduate,
post-graduate internships; formal
courses; workshops; seminars
• SERVICES: consultation, survey,
analysis (examination), documentation,
treatment

Conservation Center for Art and Historic Artifacts

264 S 23rd St., Philadelphia, PA
19103, USA; (215) 545-0613; Fax:
(215)735-9313; ccaha@shrsys.hslc.org
• PUBLICATIONS: brochures, press
releases, technical leaflets
• EDUCATION/TRAINING: workshops in
collections care, internships, lecture
series
• SERVICES: information, consultation,
survey, treatment, emergency assistance

Conservation Services

Bishop Museum, PO Box 19000,
Honolulu, HI 96817, USA;
(808)848-4112; Fax: (808)841-8968
preservation@bishop.bishop.hawaii.org
• EDUCATION/TRAINING: internships in
collections management, lectures
• SERVICES: information, consultation,
treatment

Harry Ransom Humanities Research Center

Conservation Dept., Box 7219, Austin,
TX 78713-7219, USA; (512)471-9117;
Fax: (512)471-9646

• PUBLICATIONS: newsletter (*HRHRC Bulletin*)
• EDUCATION/TRAINING: graduate and
postgraduate internships (other
internships may be arranged),
audiovisual training, workshops
• SERVICES: consultation, survey,
analysis, scientific research,
documentation, treatment

Intermuseum Conservation Association

Allen Art Building, Oberlin, OH
44074, USA; (216)775-7331;
Fax: (216)774-3431
• PUBLICATIONS: proceedings of
seminars, topical handouts
• EDUCATION/TRAINING: graduate and
postgraduate internships
• SERVICES: consultation, survey,
documentation, treatment

New York State Office of Parks, Recreation and Historic Preservation

Bureau of Historic Sites, Collections
Care Center, Peebles Island, PO Box
219, Waterford, NY 12188, USA;
(518)237-8643; Fax: (518)235-4248
• PUBLICATIONS: technical handouts,
historic preservation manual,
promotional and informational
brochures, leaflets, reports, books
• EDUCATION/TRAINING: pregraduate,
graduate, postgraduate internships;
workshops
• SERVICES: consultation, survey,
analysis, scientific research,
documentation, treatment

Northeast Document Conservation Center (NEDCC)

100 Brickstone Square, Andover, MA
01810-1494, USA; (508)470-1010;

Fax: (508)475-6021

• PUBLICATIONS: newsletter *(News)*, technical leaflets, books on paper and book conservation

• EDUCATION/TRAINING: graduate and postgraduate internships, formal courses in preservation administration, workshops, seminars

• SERVICES: consultation, survey, documentation, treatment, emergency care

Rocky Mountain Conservation Center (RMCC)

University of Denver, 2420 South University Blvd., Denver, CO 80208, USA; (303)733-2712; Fax: (303)733-2508

• PUBLICATIONS: leaflets, reports, periodic books

• EDUCATION/TRAINING: preprogram aides; pregraduate, graduate, and postgraduate internships; workshops

• SERVICES: consultation, survey, treatment, information for archeological, ethnological, manuscript, and art objects

Society for Preservation of New England Antiquities Conservation Center

185 Lyman St., Waltham, MA 02154, USA; (617)891-1985; Fax: (617)227-9204

• PUBLICATIONS: newsletter *(SPNEA News)*, journal *(Old Time New England Journal)*, house guides, program flyers, technical books

• EDUCATION/TRAINING: workshops, lecture series

• SERVICES: consultation, survey, analysis, scientific research, documentation, treatment

Williamstown Art Conservation Laboratory, Inc.

225 South St., Williamstown, MA 01267, USA; (413)458-5741; (413)458-2314

• PUBLICATIONS: leaflets, technical handouts, reports, books

• EDUCATION/TRAINING: pregraduate, graduate, postgraduate internships; formal courses; workshops; lecture series

• SERVICES: information, consultation, survey, analysis, documentation, treatment

Conservation and Preservation Institutions, Divisions, and Organizations

This general category includes government and privately funded conservation institutions as well as conservation and preservation divisions such as those operating in major museums, libraries, and archives. Generally, these organizations are involved in four major areas of activity: information dissemination, conservation treatment and collections care, research, and education. Several have sophisticated analytical equipment and scientific laboratories and maintain excellent conservation and preservation libraries and information centers. Conservation laboratories and information and outreach centers may also exist within national park service agencies as a resource for the preservation and care of both natural history and historical collections.

This category also includes preservation commissions and coalitions that promote and facilitate broad preservation goals and conservation efforts, as well as encourage the development of partnerships in the private and public sectors. Some administer projects and programs, provide grants and internships, and produce collections care publications.

Australian Institute for Conservation of Cultural Material, Inc. (AICCM)
GPO Box 1638, Canberra, ACT 2601, Australia; (06)2434 531; Fax: (06)2417 998; gina.drummond@awn.gov.au

Canadian Conservation Institute (CCI)
1030 Innes Rd., Ottawa, ON K1A 0M5, Canada; (613)998-3721; Fax: (613)998-4721
• PUBLICATIONS: newsletter *(CCI Newsletter)*, leaflets *(CCI Notes)*
• EDUCATION/TRAINING: workshops, graduate and postgraduate internships
• SERVICES: conservation research and assistance to museums (training, practice, science and technology); conservation library may be accessed through Interlibrary Loan

Canadian Museum of Nature (CMN)
Collection Division, PO Box 3443, Station D, Ottawa, ON K1P 6P4, Canada; (613) 998-5673; Fax: (613)990-7582
• EDUCATION/TRAINING: workshops on risk assessment

Commission on Preservation and Access
1400 16th St. NW, Suite 740, Washington, DC 20036-2217, USA; (202)939-3400; Fax: (202)939-3407
• PUBLICATIONS: monographs and task force reports on issues related to preservation of library and archival materials in all formats

Getty Conservation Institute (GCI)
1200 Getty Center Drive, Los Angeles, CA 90049, USA; (310)440-6900
• PUBLICATIONS: newsletter *(GCI Newsletter)*, technical report series *(Research in Conservation)*, books (e.g. *The Conservation of Artifacts Made from Plant Materials,* 1990), conference proceedings and other joint publications with professional organizations (e.g. *Art and Archaeology Technical Abstracts, International Index on Training in Conservation,* 1994)
• EDUCATION/TRAINING: courses, workshops
• SERVICES: information, Conservation Information Network (CIN)

Heritage Preservation (formerly National Institute for the Conservation of Cultural Property, Inc. [NIC])
3299 K St. NW, Suite 602, Washington, DC 20007, USA; (202)625-1495; Fax: (202)625-1485
• PUBLICATIONS: newsletters *(Heritage Preservation Update)*, study reports, books (e.g. *Preserving Natural History Collections: Chronicle of Our Environmental Heritage*, 1994; *Caring for Your Collections*, 1992; *Collections Care: A Selected Bibliography*)
• EDUCATION/TRAINING: seminars, symposia
• SERVICES: Conservation Assessment Program (CAP)

Image Permanence Institute
Rochester Institute of Technology, 70 Lomb Memorial Dr., Rochester, NY 14623-5604, USA; (716)475-5199; Fax: (716)475-7230
• EDUCATION/TRAINING: seminars and workshops on preservation of photographic materials

International Centre for the Study of the Preservation and Restoration of Cultural Property (ICCROM)
Via di San Michele 13, 00153 Rome, Italy; 39(6)585-531; Fax: 39(6)5855-3349; mc5356@mclink.it
• PUBLICATIONS: newsletter *(Newsletter-ICCROM)*, conference proceedings, directories (e.g. *International Index on Training in Conservation*, 1994, with GCI), catalogues, books, videotapes
• EDUCATION/TRAINING: courses offered by the Architecture Conservation Program, Museums and Collections Program, and the Science and Technology Program

Library of Congress
Preservation Directorate, 101 Independence Ave. SE, Washington, DC 20540, USA; (202)707-5213; Fax: (202)707-3434; nppo@loc.gov
• PUBLICATIONS: preservation leaflet series, pamphlets, monographs, audiovisual programs, fact sheets
• EDUCATION/TRAINING: advanced conservation internships and graduate conservation training programs

Museums and Galleries Commission
Conservation Unit, 16 Queen Anne's Gate, London SW1 H9AA, UK; 44(171)233-4200; Fax: 44(171)233-3686
• PUBLICATIONS: leaflets, reports, books (*Standards in the Museum Care of Biological Collections*, 1992)
• EDUCATION/TRAINING: advice on courses available in the UK, directory (*Training in Conservation*), in conjunction with UKIC
• SERVICES: data base containing information on private workshops, technical information, grants

National Archives of Canada
395 Wellington St., Ottawa, ON K1A 0N3, Canada; (613)995-5138; Conservation Dept.: (613)996-7254
• PUBLICATIONS: newsletter *(Interfax)*, journal *(The Archivist)*
• EDUCATION/TRAINING: undergraduate, graduate, postgraduate internships in archival and photographs preservation

National Library of Canada
395 Wellington St., Ottawa, ON K1A 0N4, Canada; (613)995-9481; Fax: (613)996-7941; Publications: (613)995-7969
• PUBLICATIONS: newsletter *(National*

Library News), books
• EDUCATION/TRAINING: conservation training is provided through the National Archives of Canada

National Park Service (NPS)
PO Box 37127, Washington, DC 20013, USA; Museum Management Program: (202)343-8142; Heritage Preservation Services Division: (202)343-9573; Fax: (202)343-4018
• PUBLICATIONS: preservation briefs and technical notes (*Conserv-o-gram*), magazine (*Cultural Resource Management* [CRM]), directory (*Cultural Resource Training),* technical books (e.g. *Museum Handbook* Vol. 1, 1990; Vol. 2, 1992)
• EDUCATION/TRAINING: workshops

Natural History Museum, London
Robert Huxley, Dept. of Botany, London SW7 5BD, UK
• EDUCATION/TRAINING: workshops and short courses

Parks Canada
Dept. of Canadian Heritage, Historic Resource Conservation Branch, 1800 Walkley Rd., Ottawa, ON K1A 0M5, Canada; (613)993-2125;

Fax: (613)993-9796
• PUBLICATIONS: reports
• EDUCATION/TRAINING: internships, workshops

Smithsonian Center for Materials Research and Education (SCMRE, formerly Conservation and Analytical Laboratory [CAL])
Museum Support Center, Smithsonian Institution, Washington, DC 20560, USA; (301)238-3700; Fax: (301)238-3709
• EDUCATION/TRAINING: pregraduate, undergraduate, graduate, postgraduate internships; training program in furniture conservation
• SERVICES: information, consultation

US National Archives at College Park
8601 Adelphi Rd., College Park, MD 20740-6001, USA; (301)713-6705; Fax: (301)713-6653
• PUBLICATIONS: reports, handbooks
• EDUCATION/TRAINING: courses and training in document/archival conservation
• SERVICES: funding programs for document conservation and microfilming

Conservation Education and Training Programs

The following list includes institutions, worldwide, that offer masters and PhD degrees and/or diploma programs, courses and/or workshops that especially pertain to the conservation and care of natural history collections. Undergraduate programs offer BA or BS degrees, graduate programs usually offer MA or MS degrees. A few programs offer PhDs. Internships are available at the pregraduate (i.e., pre-masters), post-graduate (i.e., post-masters), and graduate (i.e., masters candidates) level. For additional information on education and training opportunities see Pearson (1993) and AIC (1994), as well as directories listed in the previous sections under the Getty Conservation Institute and ICCROM, the Museums and Galleries Commission, and the National Park Service.

Baylor University
Dept. of Museum Studies, PO Box 97154, Waco, TX 76798-7154, USA; (817)755-1110; Fax: (817)755-1173
• EDUCATION/TRAINING: undergraduate, graduate degrees in museum studies including collections care and management

Cambridge University
Geological Conservation Unit, Cambridge CB3 0EZ, UK; 44(1223)62522; Fax: 44(1223)60779
• EDUCATION/TRAINING: graduate program in natural history conservation (under development)

Campbell Center for Historic Preservation Studies (CCHPS)
203 East Seminary, PO Box 66, Mount Carroll, IL 61053, USA; (815)244-1173
• PUBLICATIONS: course catalogues
• EDUCATION/TRAINING: courses in architectural preservation, collections care, conservation

Conservation Center, Institute of Fine Arts
New York University, 14 E. 78th St., New York, NY 10021, USA; (212)772-5800; Fax: (212)772-5807
• PUBLICATIONS: newsletter *(Conservation Center)*
• EDUCATION/TRAINING: pregraduate, graduate, postgraduate courses; internships

The George Washington University
Museum Studies Program, 801 22nd St. NW, T-215, Washington, DC 20052, USA; (202)994-7030; Fax: (202)994-7034
• PUBLICATIONS: newsletter *(Museum Studies),* course catalogues
• EDUCATION/TRAINING: graduate degree, certificate program in museum studies, including collections care administration, preventive conservation courses, workshops

Ontario Museum Association (OMA)
George Brown House, 50 Baldwin St., Toronto, ON M5T 1L4, Canada; (416)348-8672; Fax: (416)348-0438
• PUBLICATIONS: newspaper (*Currently*), journal (*Ontario Museum Annual*), course brochure
• EDUCATION/TRAINING : museum studies certificates through part-time programs; distance education certificates and seminars
• SERVICES: task forces, extensive education programs

Queen's University
Graduate Studies, Art Conservation Program, Art Centre Extension, Kingston, ON K7L 3N6, Canada; (613)545-2156; Fax: (613)545-6889
• PUBLICATIONS: course catalogues
• EDUCATION/TRAINING: graduate degree in conservation

Rochester Institute of Technology (RIT)

Technical & Education Center, 66 Lomb Memorial Drive, Rochester, NY 14623, USA; (716)475-2736; Fax: (716)475-7052
• PUBLICATIONS: brochures, technical leaflets
• EDUCATION/TRAINING: lectures, seminars, workshops

Royal Botanic Gardens, Kew

Richmond, Surrey TW9 3AB, UK; 44(181)940-1171; Fax: 44(181)332-5197
• PUBLICATIONS: alumni newsletter (*TechniQues*) for the following course
• EDUCATION/TRAINING: international diploma course in herbarium techniques (countries not previously represented are given priority)

State University College at Buffalo

Art Conservation Dept., 230 Rockwell Hall, 1300 Elmwood Ave., Buffalo, NY 14222-1095, USA; (716)878-5025; Fax: (716)878-6914
• PUBLICATIONS: newsletter (*Art Conservation Department Newsletter),* department brochure, directory
• EDUCATION/TRAINING: graduate degree in conservation
• SERVICES: consultation, conservation surveys, scientific analysis, treatment

Straus Center for Conservation

Harvard University Art Museums, 32 Quincy St., Cambridge, MA 02138, USA; (617)495-2392; Fax: (617)495-9936
• PUBLICATION: brochures, course/training catalogue, books
• EDUCATION/TRAINING: graduate, postgraduate internships

Texas Tech University

Museum Science Program, Box 43191, Lubbock, TX 79409-3191, USA; (806)742-2442; Fax: (806)742-1136
• EDUCATION/TRAINING: graduate degree in museum studies

University of Canberra

Conservation of Cultural Materials Program, PO Box 1, Belconnen, ACT 2616, Australia
• EDUCATION/TRAINING: graduate program in conservation

University of Delaware

Art Conservation Program, 303 Old College, University of Delaware, Newark, DE 19716, USA; (302)831-2479; Fax: (302)831-4330
• PUBLICATIONS: course brochures
• EDUCATION/TRAINING: undergraduate, graduate, PhD degrees in conservation

University of Göteborg

Institute of Conservation, Bastionsplatsen 2, S/411 08 Göteborg, Sweden; 46(317)734700
• PUBLICATIONS: course catalogues
• EDUCATION/TRAINING: undergraduate, postgraduate degrees in conservation, including conservation of natural science specimens

University of Leicester

105 Princess Rd. East, Leicester LE1 7L6, UK; 44 (116)533-522522
• EDUCATION/TRAINING: graduate program in museum studies

University College, London

Institute of Archaeology, 31-34 Gordon Sq., London WC1 H0PY, UK; 44(171)387-9651
• PUBLICATIONS: course catalogues, brochures

• EDUCATION/TRAINING: undergraduate conservation training program, graduate program in museum studies, courses, workshops, summer schools in archaeology and conservation

University of Nebraska—Lincoln
Museum Studies Program, 307 Morrill Hall, PO Box 880356, Lincoln, NE 68588, USA; (402)472-6465; Fax: (402)472-8899
• PUBLICATIONS: newsletter *(The Mammoth)*, journal, leaflets, reports *(Museum Notes)*, books
• EDUCATION/TRAINING: graduate courses in collections management and conservation

University of Texas at Austin
Graduate School of Library and Information Science, Preservation and Conservation Studies, SZB 564, Austin, TX 78712-1276, USA; (512)471-8290; Fax: (512)471-8285; glabs@utxdp.dp.utexas.edu
• PUBLICATIONS: newsletter *(Conservation and Administration News)*

• EDUCATION/TRAINING: graduate degree in library and archives conservation and administration

University of Toronto
Museums Studies Program, Robarts Library, Suite 6003, Toronto, ON M5S 1A1, Canada; (416)978-4211; Fax: (416)978-8821
• PUBLICATIONS: course catalogues
• EDUCATION/TRAINING: graduate degree in museum studies

University of Victoria
Cultural Resource Management Program, Division of Continuing Studies, PO Box 3030, Victoria, BC V8W 3N6, Canada; (604)721-8462; Fax: (604)721-8774
• PUBLICATIONS: course catalogues
• EDUCATION/TRAINING: undergraduate and professional development courses in museum studies, heritage conservation, cultural management, diploma program in cultural conservation

A Partial Listing of Organizations with Conservation/Preservation Interests

The following list includes collections-oriented organizations that either serve the herbarium community directly or offer relevant literature or other resources. Notable among them is the Association for Systematics Collections (ASC) which is composed of eighty-three North American institutions that maintain systematics collections. Working with Heritage Preservation (formerly NIC) and SPNHC, the ASC has organized symposia and conferences, conducted studies on natural science collections needs, and developed collections policy guidelines (Hoagland, 1994).

American Association of Museums (AAM)

1575 Eye St. NW, Suite 400,
Washington, DC 20005, USA;
(202)289-1818; Fax:(202)289-6578
• PUBLICATIONS: newsletters (*AVISO, AAM Network News)*, magazine (*Museum News)*, catalogue (*AAM's Bookstore for Museum Professionals)*, books (e.g. *Official Museum Directory)*, pamphlets, reports, leaflets, reprints
• EDUCATION/ TRAINING: professional education seminar series, workshops, annual meeting
• OTHER: standing professional committees, professional interest committees, task forces, councils

American Library Association (ALA)
Association for Library Collections and Technical Services (ALCTS)

Resources and Technical Services Division, 1301 Pennsylvania Ave., Suite 403, Washington, DC 20004, USA; (202)628-8410; Fax: (202)628-8419; alawash@alawash.org
• PUBLICATIONS: newsletter (*ALCTS Newsletter)*, journal (*Library Resources* and *Technical Services)*, directories, bibliographies, special publications (*American Libraries, Booklist, Choice, Book Links)*
• EDUCATION/TRAINING: library and information studies accreditation, coordinates pre-conference/conference programs, regional divisions of the organization also offer conferences and workshops

Association of Research Libraries (ARL)

21 Dupont Circle, Suite 800,
Washington, DC 20036, USA;
(202)296-2296; Fax: (202)872-0884;
arlq@cni.org
• PUBLICATIONS: newsletter (*ARL Bimonthly Newsletter)*, books
• EDUCATION/TRAINING: seminars, workshops, conferences

Association for Systematics Collections (ASC)

1725 K St. NW, Suite 601,
Washington, DC 20006, USA;
(202)835-9050; Fax:(202) 835-7334
• PUBLICATIONS: newsletter (*ASC Newsletter)*, books
• EDUCATION/TRAINING: workshops, conferences
• SERVICES: develops, funds, and implements projects and programs with government and private organizations

Biology Curators Group (BCG)

c/o Kathie Way, Zoology Dept., The Natural History Museum, Cromwell Rd., London SW7 5BD, UK
• PUBLICATIONS: newsletter (*The Biology Curator)*; book (*Manual of Natural History Curatorship)*
• EDUCATION/TRAINING: conferences and workshops on collections care

International Council of Museums (ICOM) Canada

#400, 280 Metcalfe St., Ottawa, ON K2P 1R7, Canada; (613)567-0099; Fax: (613)233-5438
• PUBLICATIONS: newsletter (*ICOM Canada Newsletter)*, journal (*UNESCO Museum Quarterly)*
• SERVICES: annual meetings in conjunction with CMA

International Council of
Museums (ICOM) (United States
National Committee) – join
through American Association of
Museums (AAM)

Institute of Paper Science and
Technology
American Museum of Paper Making,
500 10th St. NW, Atlanta, GA 30318,
USA; (404)894-7840; Fax: (404)894-4778
• Publications: brochures, bulletins,
books, multimedia CD-ROMs
• Education/Training: graduate and
PhD programs, research and testing

Society of American Archivists
(SAA)
600 S Federal St., Suite 504, Chicago,
IL 60605, USA; (312)922-0140;
Fax: (312) 347-1452

• Publications: newsletter *(Outlook)*,
journal *(The American Archivist)*,
pamphlets, manuals, books
• Education/Training: annual
meetings, workshops, training in
archival practices and theory

SOLINET
Southeastern Library Network Inc.,
Suite 200, 1438 West Peachtree St.,
Atlanta, GA 30309, USA;
(404)892-0943; (800)999-8558;
Fax: (404)892-7879;
solinet_information@solinet.net
• Publications: monographs on
preservation topics, informational
handouts on supply sources,
procedures, guidelines
• Education/Training: workshops
• Services: reference, Audiovisual Loan
Service, consultation, microfilm

Conservation Abstracts and Selected Conservation Publishers

It is important for the conservation professional to keep abreast
of the wealth of information that is offered in the literature. In addition to the resources listed below Torres (1990a, 1990b) and
Duckworth et al. (1993) provide useful information and bibliographies on natural history collections care.

Abbey Publications, Inc.
7105 Geneva Dr., Austin, TX 78723,
USA; (512)929-3992; Fax: (512)929-3995
• Publications: newsletters *(The Abbey
Newsletter: Bookbinding and
Conservation, The Alkaline Paper
Advocate)*
• Services: answers inquiries, keeps
databases of alkaline mills and
permanent papers

Archetype Books
31-34 Gordon Sq., London WC1

H0PY, UK; 44(171)380-0800;
Fax: 44(171)380-0500
• Publications: conservation, museum
studies, collections care books

Art and Archaeology Technical
Abstracts
Getty Conservation Institute, 1200
Getty Center Drive, Los Angeles, CA
90049-1684, USA; (310)440-6900;
Fax: (310)440-7702
• Publications: voluminous
compilation of over 3000 abstracts on

conservation, materials, fabrication, technology, analysis, deterioration, and other related topics (published semiannually and available on-line through the bibliographic database [BCIN] of the Conservation Information Network [CIN])

Butterworth-Heinemann
225 Wildwood Ave., Unit B, Woburn, MA 01801, USA; (800)366-2665; Fax: (800)446-6520; orders@BH.com
• PUBLICATIONS: conservation and related technical publications

Technology and Conservation®️ of Art, Architecture, and Antiquities
Susan Schur, Editor, One Emerson Place, Boston, MA 02114, USA; (617)623-4488
• PUBLICATIONS: magazine on conservation and technology topics, frequently contains extensive bibliographies (published periodically)

Relevant Botanical Publications

• *The Herbarium Handbook*
Bridson and Forman (eds), 1992 (revised edition); Forman and Bridson (eds), 1989; published by Royal Botanic Gardens, Kew
A general overview of herbarium curation; the revised edition is more "conservation-oriented" than the first.

• **Herbarium News**
Newsletter containing information and news of interest to herbarium personnel including: notes on curation, announcements of meetings, job advertisements, deaths, staff changes, temporary closures, new publications, etc. Published by Missouri Botanical Garden, PO Box 299, St. Louis, MO 63166-0299, USA;(314)577-5100; Fax: (314)577-9595; http://hoya.mobot. org/mobot/research/herbarium/

• *Index Herbariorum*
Holmgren, Holmgren, and Barnett (eds), 1990; published for the International Association for Plant Taxonomy (IAPT) by New York Botanical Garden. A detailed directory of the public herbaria of the world; can be searched through the New York Botanical Garden web site: http://www.nybg.org/bsci/ih/ih.html

• **Taxon**
Journal of the IAPT devoted to systematics and evolutionary biology with emphasis on botany. Articles on herbaria are common; regular columns on collections and collectors, herbaria and institutions, news and notes. To subscribe: IAPT, c/o Dept. of Botany, Smithsonian Institution, Washington, DC 20560, USA

Online Conservation and Herbarium Resources

Online conservation services are growing rapidly and the list continually expands. Many of the organizations listed in this paper also have home pages which contain information on their purpose and offerings. Many listservers announce upcoming conferences, meetings, or courses.

Conservation DistList
whenry@lindy.stanford.edu;
(415)725-1140
• Services: discussion group on
conservation issues

Artefacts Canada (formerly Conservation Information Network [CIN]) (a collaborative venture of CCI, CHIN, GCI, SCMRE, ICCROM, ICOMOS, ICOM)
Client Services, Artefacts Canada, Communications Canada, Terrasses de la Chaudière, 15 Eddy St., Hull, PQ J8X 4B3, Canada; (800)520-2446; Fax: (819)994-9555
• Services: online database and electronic mail system; access to the conservation bibliographic database BCIN (which contains the AATA abstracts); information on materials and suppliers

Conservation OnLine (CoOL)
http://pelimpsest.stanford.edu
• Services: information server for conservation professionals, contains information on many organizations listed in this section

IMPNW (formerly HERB-L)
IMPNW@idbsu.idbsu.edu
• Services: list for botanists with an interest in taxonomy/systematics of the intermountain and pacific northwest region of the United States and adjacent Canada

Materials-L
listproc@williams.edu
• Services: listserver to facilitate discussion about materials use and testing

Museum-L
listserv@unmvma.unm.edu
• Services: facilitates discussion about materials used for storage, exhibition, conservation

NHCOLL-L
To subscribe, email to:
listproc@lists.yale.edu
Type: subscribe nhcoll-l followed by full name
• Services: unmoderated list sponsored by ASC and SPNHC, open to all individuals interested in the management and care of natural history collections

South African Herbaria
http://www.ru.ac.za/departments/herbarium/SAHWG/address.html
• Services: lists email addresses and contact persons for herbaria in Southern Africa

TAXACOM
TAXACOM@cmsa.berkeley.edu
• Services: A discussion list for biological systematics. Topics include systematics, taxonomy, nomenclature, issues relating to the role of, and care of, herbaria and other natural history collections, databasing, electronic publications

Taxonomists-ONLINE
gopher://gopher.unm.edu:70/11/
academic/biology/pto

• SERVICES: lists many herbaria, and hundreds of systematists worldwide including address, email, phone, fax, and academic or collections interests

Conservation Funding Organizations

Most of the following organizations provide funds for collection improvement programs that involve assessing the needs of collections, upgrading storage equipment and supplies, and rehousing collections. Some funding organizations sponsor symposia, workshops, and publications. Reger (1995) provides additional information on how best to tap into this private sector support.

The Bay Foundation
17 West 94th St., New York, NY
10025, USA; (212) 663-1115
• SERVICES: funding for training programs and workshops; charitable endowment with grants: for children's services, to preserve biological diversity, to support programs of native Americans, for the care of cultural and natural history collections

Canadian Museums Association (CMA)
280 Metcalfe St., Suite 400, Ottawa, ON K2P 1R7, Canada; (613)567-0099; Fax: (613)233-5438
• PUBLICATIONS: magazine (*Museogramme*), reports (*ND Report*), journal (*Muse*)
• SERVICES: funding for continuing professional development and better administration of museums, awards (including Outstanding Achievement Awards), grants, scholarships

Cultural Human Resources Council (CHRC)
189 Laurier Ave E, Ottawa, ON K1N 6P1, Canada; (613)565-7956;

Fax: (613)565-7022;
CHRC@magl.com
• SERVICES: funding program, Training Initiatives Program (TIP), provided by Human Resources Development Canada. TIP provides training and professional development, and mentorship and internship for artists and cultural workers

Institute of Museum and Library Services (IMLS)
1100 Pennsylvania Ave. NW, Room 609, Washington, DC 20506, USA; (202)606-8539
• SERVICES: funding for conservation programs and education (course work, refresher courses, seminars, conferences, internships, summer work projects, training programs)

Museum Assistance Program
15 Eddy Street, Rm 38, 3rd floor, Hull, PQ J8X 4B3, Canada; (819)997-8816
Fax: (819)997-8756
• SERVICES: provides financial and technical assistance to Canadian museums and related institutions for museum activities that support the

Museums and Galleries Commission
Conservation Unit, 16 Queen Anne's Gate, London SW1 H9AA, UK; 44(171)233-4200;   44(171)233-3686
• PUBLICATIONS: leaflets, reports, books
• SERVICES: advice on courses available in the UK, data base containing information on private workshops, technical information, grants

National Science Foundation (NSF)
4201 Wilson Blvd., Suite 1205, Arlington, VA 22230, USA; (703)306-1234; Fax: (703)306-0202;

programs: Anthropology: (703)306-1758; Biology: (703)306-1481
• SERVICES: grant programs funding conservation activities including planning, storage supplies and equipment, environmental equipment, treatment

National Endowment for the Humanities (NEH)
Office of Preservation, Room 802, 1100 Pennsylvania Ave. NW, Washington, DC 20506, USA; (202)606-8570
• SERVICES: funding for course work and training programs, grants for collections care and training

Standards and Testing Organizations

In selecting materials for conservation, it is useful to be aware of the material testing organizations and the standards that apply to conservation quality materials. This information can be used in initial investigations of appropriate storage materials for herbarium collections. For example, paper and photographic materials, in particular, have been extensively tested.

American National Standards Institute (ANSI)
11 W 42nd St., New York, NY 10036, USA; (212)642-4900; Fax: (212)398-0023
• PUBLICATIONS: newsletter *(ANSI Reporter* including ANSI Standard Action — standards currently being developed)*, standards

American Society for Testing and Materials (ASTM)
100 Barr Harbor, West Conshohocken, PA 19428, USA; (610)832-9500; Fax: (610)832-9555; service@local.astm.org

• PUBLICATIONS: informational brochures, standards, special technical publications
• EDUCATION/TRAINING: technical and professional training courses in the use and application of standards

National Information Standards Organization (NISO)
4733 Bethesda Ave., Suite 300, Bethesda, MD 20814, USA; (301)654-2512; Fax: (301)654-1721; nisohq@cni.org
• PUBLICATIONS: newsletter *(Information Standards Quarterly)*, books

Technical Association of the Pulp and Paper Industry (TAPPI)
15 Technology Park S, PO Box 105113, Atlanta, GA 30348, USA; (770)446-1400; Fax: (770)446-6947
• PUBLICATIONS: catalogue *(TAPPI)*, journal *(Tappi Journal)*, reference publications, educational books, videotapes
• EDUCATION/TRAINING: videotape/multimedia training, short courses, meetings

Health and Safety Resources

Arts, Crafts, and Theater Safety (ACTS)
181 Thompson St., #23, New York, NY 10012, USA; (212)777-0062; Fax: (212)777-0062; 75054.2542@compuserve.com
• PUBLICATIONS: newsletter *(ACTS FACTS)*, data sheets on conservation topics, instructional videotape *(First Steps)*

Center for Safety in the Arts at NYFA
1555 Ave. of the Americas, 14th floor, New York, NY 10013, USA; Voice mail: (212) 366-6900, ext. 333; csa@tmn.com
• PUBLICATIONS: newsletter *(Art Hazard News)*

CONCLUSION

Research, management, and conservation must work in concert to best serve the long-term interests of natural history collections and to insure their continued viability. A good understanding of current conservation principles, practices, and approaches provides the means to achieve this goal. The use of available knowledge and technology offered through publications, education forums, and conferences can avoid costly mistakes and the duplication of efforts, and will assist in preserving valuable collections for the future.

LITERATURE CITED

[AIC] AMERICAN INSTITUTE FOR CONSERVATION. 1994. *Conservation Training in the United States.* American Institute for Conservation of Historic and Artistic Works, Washington, DC.

[AIC] AMERICAN INSTITUTE FOR CONSERVATION. 1997. Regional Conservation Associations in the United States. AIC Newsletter 22(1)1-3.

History Collections: Chronicle of Our Environmental Heritage. National Institute for Conservation, Washington, DC.

HOAGLAND, E., editor. 1994. *Guidelines for Institutional Policies and Planning in Natural History Collections.* Association of Systematic Collections, Washington, DC.

PEARSON, C. 1993. Resources available for education and training in conservation. Pp. 231-239 in C. L. Rose, S. Williams and J. Gisbert, editors. *Current Issues, Initiatives, and Future Directions for the Preservation and Conservation of Natural History Collections.* International Symposium and First World Congress on the Preservation and Conservation of Natural History Collections. Dirección General de Bellas Artes y Archivos, Ministerio de Cultura, Madrid, Spain, vol. 3, 439 pp.

REGER, L. 1995. Collections care: catalyst for funding. Pp. 411-419 in C. L. Rose, C. A. Hawks and H. H. Genoways, editors. *Storage of Natural History Collections: A Preventive Conservation Approach.* Society for the Preservation of Natural History Collections, Iowa City, IA.

ROSE, C. L. 1993. Conservation and preservation of natural history collections in North America. Pp. 199-216 in C. L. Rose, S. Williams and J. Gisbert, editors. *Current Issues, Initiatives and Future Directions for the Preservation and Conservation of Natural History Collections.* International Symposium and First World Congress on the Preservation and Conservation of Natural History Collections. Dirección General de Bellas Artes y Archivos, Ministerio de Cultura, Madrid, Spain, vol. 3, 439 pp.

ROSE, C. L. AND A. R. DE TORRES, editors. 1992. *Storage of Natural History Collections: Ideas and Practical Solutions.* Society for the Preservation of Natural History Collections, Iowa City, IA.

ROSE, C.L., C.A. HAWKS, AND H. GENOWAYS, editors. 1995. *Storage of Natural History Collections: A Preventive Conservation Approach.* Society for the Preservation of Natural History Collections, Iowa City, IA.

TORRES, A.R. DE, editor. 1990a. *Collections Care: A Basic Reference Shelflist.* National Institute for the Conservation of Cultural Property, Washington, DC.

TORRES, A.R. DE, editor. 1990b. *Collections Care: A Selected Bibliography.* National Institute for the Conservation of Cultural Property, Washington, DC.

WILLIAMS, S. AND P. CATO. 1995. Interaction of research, management, and conservation for serving the long-term interests of natural history collections. Collection Forum 11(1):16-27.

CHAPTER 3

A Healthy Dose of the Past:
A Future Direction in
Herbarium Pest Control?

TOM J.K. STRANG

Abstract.—The modern control of pests in collections stresses less dependancy on a single-chemical line of defence, often replacing these chemicals with a choice of thermal or controlled-atmosphere fumigation techniques. Moreover, Integrated Pest Management (IPM) has expanded in collections to include environmental and structural controls that resist insect build-up in collection areas. None of these elements is particularly new; however, this time around, we have a wider range of tools to apply in concert with a more disciplined approach to collection care. This paper examines the earlier use of IPM by E.D. Merrill and compares it to the author's systematic approach to IPM in collections.

INTRODUCTION

Fifty years ago in his paper "On the control of destructive insects in the herbarium," E. D. Merrill critiqued the very system of pest control with which most of us are familiar: repellent-soaked, cabinet-housed collections.

In herbarium practice, from a historical standpoint, controls have taken the form of tight cabinets, sometimes difficult to attain when wood is used as is generally the case; by actual poisoning of the specimens; and by fumigation. In some institutions repellents are used to a certain degree. Any experienced curator will realize that such controls are only in part effective, for there is always the human element to consider. To keep any herbarium free from the depredations of the herbarium beetle and other destructive insect pests involves constant attention. Not infrequently, in little consulted material, the actual destruction of important specimens may reach distinctly large proportions before an infestation is detected. (Merrill, 1948)

Merrill condemned collection-wide poisoning and fumigating of materials as inefficient, since he noted the toxins eventually failed to protect specimens and imparted a "false sense of security" that outlasted the effect of the poison. This comment came not from a cultural fear of chemical use but from direct experience of failure of whole building fumigation with cyanide gas, repeat fumigations, and insect depredation of poisoned specimens (Merrill, 1948).

In this paper I have laid out a brief synopsis of the key points in Merrill's paper which describes a very successful pest control program for a large, growing herbarium. I then illustrate a broader pest control program that I have organized in a methodical and structured manner following the Framework for Preservation of Museum Collections developed by the Canadian Conservation Institute (CCI, 1994; Rose and Hawks, 1995) and finally, illustrate the parallel between Merrill's choice of methods and the major elements in the CCI framework.

A SYNOPSIS OF E. D. MERRILL'S
PEST MANAGEMENT SYSTEM

E. D. Merrill was critical of every element he used to control pests — and I feel would not weight any procedure as 100% effective — yet he held up his systematic approach as very effective indeed. This is heartening as the greatest criticism of any pest control program being adopted after the loss of an institution's major chemical pesticide/fumigant is that the successor system must be too "soft" on the insects now that the "hard" method is unavailable. When I began my work on pest control for collections,

it was my personal opinion that a layered system of mostly efficient deterrents can be as good as, or even better than, reliance on
one control method, such as a chemical. Merrill's paper chronicled
a necessary test of that premise and the proof that such a layered
control strategy can work.

Merrill's Chemical Barrier

Naphthalene was a widely used chemical available to Merrill,
and its use can be traced to Scott et al. (1918) who provided efficacy data as newly required by the United States' Insecticide Act of
1910. Closed tulipwood containers containing flannel were salted
with *Tineola bisselliella* (Hummel), webbing clothes moth, adults
and larvae at a rate of 10 to 20 each week for several weeks. The
authors reported complete kill in the naphthalene-treated boxes
although some damage from larvae was seen in certain tests. Minor survival of *Attagenus piceus* Oliv. (now *Attagenus unicolor*
Brahm, carpet beetle) was seen. All the untreated boxes showed
insect damage, survival, and breeding. While this was not stun-
ning efficacy, naphthalene was widely popularized and used.

By 1948, Merrill noted objectionable and allergic reaction by
individuals to naphthalene, but still argued that naphthalene use
was reasonable. Even so, he only considered naphthalene to be a
repellant and an ineffective fumigant which did not achieve lethal
concentrations inside the cabinets. As a repellant, naphthalene
only encouraged invading insects to preferentially colonize untreated piles of delectable foliage that had been systematically laid
down by Merrill throughout the facility as traps.

By 1992 the National Toxicology Program's investigation of
naphthalene-exposed mice had documented lesions, chronic respiratory inflammation, and "a marginal increase of neoplasms
(malignant, benign, or combined) in which the strength of the response is less than that required for clear evidence" (NTP, 1992).

As with naphthalene, most of the toxins applied to natural history collections have come under review, restriction, or downright
ban. Since collections are a minor market for pesticides, we have
to consider alternate strategies in order to maintain control over
the future preservation of our collections.

Merrill's Physical Barrier

The protective ability of "tight steel cases", containers introduced to herbaria in 1905, was also described as uncertain since opportunities for insects to enter the cases are provided by leaving doors open longer than required (Merrill, 1948).

In my own experience, failing to maintain intact gaskets is an "old saw" in collection care, but old saws are often dull implements, and one still finds inadequate seals due to faulty maintenance. A further caution: a new case may be delivered with holes at folded metal corners, or other compromising perforations, that are never noticed unless the case is inverted and inspected. Simple daubs of sealant will prevent insects that are sheltering between the case and floor, or between the walls of double walled cases, from recolonizing the case proper. There are many styles of cases used in museums with underside openings ranging from a millimetre to several centimetres. The projecting wall structure which acts as the foot for the case eventually directs prospecting insects toward the corner perforation.

Merrill's Attractant Insect Traps

Merrill's list of plants attractive to *Lasioderma serricorne* (Fab.), the cigarette beetle (Table 1), is useful to collection managers as it can be used to focus the limited time available for inspection on likely targets of attack. Experienced staff are often quite aware of particularly sensitive specimens in their collection and focus inspections on these.

TABLE 1. List of plants that are attractive to *Lasioderma serricorne* (Fab.), the cigarette beetle (Merrill, 1948).

Caprifoliaceae	Capparidaceae	Ranunculaceae	Umbelliferae
Cruciferae	Nymphaeaceae	Araliaceae	Asclepiadaceae
Liliaceae	Labiatae	Apocynaceae	Araceae
Compositae	Moringaceae	some Leguminosae	Solanaceae
Tropaeolaceae	some Rosaceae	Scrophulariaceae	Papaveraceae

Merrill advocated the use of "bug traps" consisting of twelve, six inch thick bundles containing a selection of attractive plants: *Asclepias* (Asclepiadaceae), *Apocynum* (Apocynaceae), *Sonchus* and

Taraxacum (Compositae), *Nicotiana* (Solanaceae), Umbelliferae, Cruciferae, Papaveraceae, Labiatae, Scrophulariaceae. The bundles were dated, placed throughout the collection space, fumigated quarterly, re-dated, and redistributed.

Merrill's Appraisal of His System

I believe that our experience over a decade, with not a single live beetle, pupa, or grub reported in the last eight years, rather definitely demonstrates the efficacy of this very simple and relatively inexpensive method of eliminating these herbarium pests. The scheme is practicable, efficient and what is more, inexpensive; in fact much less expensive in time and in material than the current practice, in some large herbaria, of poisoning all material with corrosive sublimate before the mounted sheets are distributed into the herbarium, to say nothing of repeated fumigations. Perhaps the very simplicity of the plan will militate against its general use, but definitely it is worth a thorough trial (Merrill, 1948).

The Arnold Arboretum Herbarium where Merrill worked was greater than 700,000 sheets at the time he was writing, with nearly 500,000 sheets having been loaned or acquired in one decade. Merrill's description implies a period of two or more years for the full benefit of changes in pest control strategy to accumulate.

It is an unfortunate irony that the cardboard box devised by Merrill to store herbarium sheets became known as "Merrill's Perils", out of the sad experience of herbaria leaving them unprotected on top of herbarium cases for indefinite lengths of time, resulting in a high rate of infestation (J. Solomon, Missouri Botanical Garden, pers. comm.). This, as with his assessment of steel containers, is reflected in the comment:

I have noted above that there is always the human element to be considered. Unless curators assume their proper responsibilities and keep constantly on the alert for indications of the presence of the herbarium beetles, no matter what system be followed reinfestation may occur at any time during the breeding season of the insects involved. Proper precautions and eternal vigilance on the part of responsible officials is the price that an insect free herbarium exacts. The cooperation of all staff members and of others who have occasion to consult stored material is important,

and this whether the specimens be poisoned or not. Whether an infestation be light or severe, it should be attended to immediately once it is
discovered. To delay in applying control methods is merely to invite disaster in the form of the destruction of often very valuable and utterly irreplaceable specimens. (Merrill, 1948)

Merrill has described more than the common pest control system we have often inherited. It is rare to encounter anyone using
bundles of plants to drain off potential intruders (unless it is an
open cabinet filled with *Letuca).* In fact, until the relatively recent
and growing emphasis on using adhesive traps and the development of pheromone traps for a few herbarium pests, Merrill's system was certainly too simple to be generally used. The advantage
of combining an attractant with an adhesive trap is the ease of
counting and identifying pests. It is unclear whether Merrill
counted the numbers of pests, as we would advocate today in order to
create statistics on our control strategies and guide our programs.

A WIDER VIEW —
A SYSTEMATIC PRESENTATION ON
INTEGRATED PEST MANAGEMENT

Merrill established a layered pest control strategy which gave
results that were superior to a strategy of sole reliance on a chemical to protect everything. It is perhaps understandable that most
collections devolved to the one chemical approach given the enthusiasm of the last fifty years for pesticides based on their immediate effect. It is, however, short-sighted to base one's pest control
program on one chemical given increasing insect resistance and
regulatory changes.

Integrating pest management into a museum's mandate and
operations is essential for the long-term preservation of the stored
artifacts or specimens. A pest management program is successfully
integrated when its requirements are not seen as impositions or
crisis driven, but as ongoing and routine functions in the collection.

A pest management program starts with a survey of the present
system and follows with a design that corrects shortcomings. Implementation should be carried out by an individual or group that

has the authority to achieve improvements within the museum system.

An integrated pest management (IPM) program is developed for a specific industry or site to reduce pest problems to a tolerable level. IPM relies on effective sanitation, identifying and monitoring of pests, and strategic use of controls from a range of environmental, biological, and chemical agents. Commercial IPM programs aim to reduce pest problems to an action threshold that is decided by cost against return. Within museums, the tolerable level of damage within a collection may be very small due to the unique and valuable nature of the stored items. The risk may be very large as most organic materials are prone to depredation by pests.

The applicability of a control method depends on extent of infestation, infesting organism, types of material affected, deleterious side effects of the control method, and immediate cost. The choice of control should be made by people who are aware of these elements. Typically, collection care and conservation staff, and licensed pest control operators with broad museum experience are able to achieve this balance because of the technical knowledge they possess and their interest in the welfare of the stored material and of the collection's handlers.

IPM programs stress reduced pesticide use for health, environmental, and technical reasons. The goal is to maintain control, reduce harmful, cumulative exposure of people to pesticides, and preserve pesticide efficacy by lowering the rate of increase of pesticide-resistant strains. This is only possible with an effective monitoring program and knowledge of the pest organisms. Continual monitoring allows timely use of pesticides, reduces the total volume applied, and increases the effectiveness of non-pesticide approaches.

The IPM Survey

The pest management survey is intended to describe the existing system. This involves an assessment of the three means by which museums control damaging agents (CCI, 1994; Rose and Hawks, 1995):

a) the building and surroundings,

b) portable fittings and hardware,

c) staff and procedures.

Analysing these means for their positive and negative influence on the ability to combat the damaging effects of pests reveals strengths and weaknesses in the museum's system of pest management.

A baseline survey of pest distribution and population is useful for revealing the magnitude of the current problem, identifying deleterious agents, and allowing future measurement of progress. Often, people do not want to do this step; they only want to remove the offending organisms in an expedient manner. As a survey method coordinated trapping is moderately easy to institute, only risks offending those few who like to pretend they don't have "critters", and provides a measure of progress which can be extremely valuable, especially if resources get scarce and one has to constantly re-justify allocations.

Faults at the building level are often a matter of maintenance or upgrading of seal details, eliminating external habitats and other attractors, and establishing defined quarantine and control facilities. *Faults at the portable fitting and hardware level* are often a lack of protective containers for specimens and deteriorated seals on cases, aggravated by crowding and disorder that hinders detection and control. *Faults at the procedures level* are often poor sanitation, no efforts toward early detection, crisis driven response, disinterest, or lack of coordination between jurisdictions.

The state and interaction of these three means will govern the effectiveness of a pest management program. Assessing the existing system, its underlying assumptions and constraints provides a means to plan and effect improvement.

Design of an IPM Program

The long-term ability of an institution to protect its collection can only be improved when significant shortcomings are identified. Improvement requires coordination between people carrying out collection management, conservation, pest control operations, curatorial, janitorial, engineering, planning, and funding roles. Having representatives from these areas on, or in regular contact with, a pest management committee is recommended.

This paper expands on the pest control section of the Framework for Preservation for Museum Collections (CCI, 1994) which includes details for the design of an IPM program that improve the means (buildings, fittings, procedures) and stages (avoid, block, detect, respond, recover) of pest control, and reflect the institution's function, physical layout, and resources. The listed control details often extend to the reduction of risk from other agents of deterioration, particularly in the area of blocking or slowing their spread. Consult the CCI framework to gain an appreciation of this larger scale integration. While the argument for spending time and money on a detail may primarily be for pest control, simultaneously reducing problems from other agents strengthens the reasons behind the design.

Three Means of Control

The individual who is responsible for pest control should direct and plan pest management activities on three means of control: building, portable fittings, and procedures. There are different degrees of cost, portability, supervision, and coordination associated with each of these means. The possibility of solving problems by all three means should be considered before adopting any one.

1. Building

When choosing a building or building site for a collection, it is important to be aware of the past and present operations on neighbouring sites, and those of previous tenants on the site under consideration, as they may contribute high numbers of pests.

The building is the largest enclosure for a collection. Without it, you can imagine the collection sitting out of doors deteriorating at rates many thousands of times faster than within the building. The building's protective ability should be fully exploited as it is both the first, and thus the defining boundary for the collection, and, short of the collection's intrinsic value, is likely the most expensive element. To this end, it is especially important to have mechanisms that will eliminate rapid threats to the building and collection, such as fire suppression systems (see Ertter, Chapter 6, this volume; Lull and Moore, this volume; CCI [1994] for more details). From the perspective of pest control, the building must function as a fairly efficient filter for those pests wishing to enter, and as an inhospitable place for those who make it inside.

2. Portable Fittings and Hardware

This level of control includes all of the building's contents used in day to day tasks. Is monitoring/trapping equipment available and being used? Are the records organised into a useful periodic report? Is specimen containment high with vulnerable specimens housed inside cases, boxes, vials, or polymer bags? Have staff reduced pest habitat by eliminating disorder and overcrowding, and using covered waste bins? Have access for inspection and sanitation activities been designed into fittings, with shelving and cases supported off the floor to allow effective sanitation underneath and to serve as a deterrent to flood damage (Williams and McLaren, 1990)? Is equipment available to combat an infestation: freezers, fumigation equipment, polyethylene or polyester sheeting to bag and quarantine objects?

3. Procedures to be carried out by staff

Is effective sanitation being performed? Has a quarantine policy been established and followed, how are exceptions handled? Is there an active monitoring program, with equally effective synthesis of the information? Are pest control precautions and procedures written into institutional policy to ensure the continued safety of both the acquired object and the objects it is housed with? Is there someone charged with pest management duties, and do they have a building-wide mandate? Are staff trained in pest eradication methods, are the techniques used routinely, and are the required materials on hand? Does the institution contract a pest control company educated in museum work for pesticide application or other services?

Five Stages of Control

A layered pest control strategy is organized around five stages of control: avoid, block, detect, respond, recover (CCI, 1994; Rose and Hawks, 1995). These stages define the actions that preserve collections from pest damage. Their order presents a hierarchy of activity that can increase in effort and cost if previous activities are not incorporated. What one cannot avoid, one tries to block. One can only respond to whatever one detects. Without avoidance, blocking, and detection, the response stage will often be a crisis management activity that summarily redirects or drains the institution's resources. The recovery stage is a tacit admission of failure to prevent damage. If you are spending a lot of time darning new holes in the objects, you are plugging away at the wrong holes in the pest management system.

1. Avoid pest attractors

- *In outside locations:*

  The waste dumpster/compactor should be vermin-proof; an outdoor yet enclosed location is preferable to placement inside a loading bay. Garbage must be covered and spills routinely washed away.

  Exterior lighting designs that reduce ultraviolet light shining on the building at night by using high-pressure sodium lamps, while using ultraviolet-rich mercury sources to draw insects away from openings such as doorways, can lower insect pressures on the building. Note that security lighting requirements (high illumination into the eyes of an intruder, good colour rendition) take precedence over insect control, so positioning and type of light source may be less than satisfactory from a pest control viewpoint.

  Ensure that water drains away from the building, especially in times of heavy precipitation. Water will not only destroy the building, but also greatly increase the incidence of mould damage to the collections, and provide water for pests.

  Remove bird, mammal, and insect nests, as these house damaging insect pests.

  Reduce floral plantings near the building as they attract adult dermestids. Bushy plants adjacent to buildings also provide cover for rodents.

  Establish a sanitary perimeter around the exterior walls of a building that is free of cover. A three foot wide pea-gravel border can reduce colonization by rodents. The sanitary perimeter not only lowers pest pressure on the structure, but also provides clear lines of sight for security, and easy access for building inspection and maintenance.

- *In inside locations:*

Remove or contain garbage and clean garbage storage areas regularly. A cool room for garbage storage will reduce odour.

Internal light traps should draw insects back toward exterior doors, away from collections.

Allow food only in specified cleaned locations, and keep it properly stored in closed containers.

Bouquets of flowers, especially wild flowers which can harbour dermestid adults, should either be inspected for insects prior to entry or banned altogether.

Replace adhesive insect and rodent traps regularly, inspect bait stations for spillage or tampering, remove snap traps and dispose of corpses quickly.

Choose structural shapes and assemblies with the intent to reduce dust accumulation or insect habitat in hard to clean areas.

Monitor sanitation of collections and offices, and ensure periodic cleaning of rarely used areas.

Establish and maintain a sanitary perimeter, that is wide enough to walk in, along the inside of exterior walls. Although it reduces "useful"space in which to cram objects against walls, a gap between shelves and walls will allow proper inspection of ceiling/wall/floor junctions for signs of pests and leaks. A clear interior line of sight allows easier security and the use, for instance, of infrared beam intrusion detection down an entire wall. Most water leaks flow down walls; most pests follow walls; upon reaching a wall in an emergency, people deserve a clear path to the escape doors. For all of these reasons, cramming in one extra shelf may not be beneficial to the collections overall.

2. Block pests

- *Exclusion by the building:*

Ensure that the building is well sealed and maintained; screen vents to prevent pest access. Construction details such as gaps at ceiling/wall/floor junctions, pipe runs, doors, and windows are flaws which can be exploited by pests. While 5 mm holes or gaps are restrictive to mice, one has to get down to 0.3 mm cracks to be largely restrictive to insect pests. A nominal 1 mm gap will be effective for keeping out some, but not all, insect species stages that damage collections. Mechanisms should be established to ensure the timely repair of flaws in the building's structure. High quality repairs are essential to lowering pest control workload. It is far better to caulk a seam once and remove an insect harbourage, than to leave the harbourage and return every few months to spray with pesticide.

- *Exclusion by portable fittings:*
Place objects in sealed storage furniture, boxes, vials, or bags.
As with building details, strive to restrict gaps to 0.3 mm to control passage of damaging insects, although many species are repelled by holes just under 1 mm. Restrict gaps to less than 5 mm to control passage of rodents. Enclosures with cracks less than 1 mm will reduce contaminant infiltration and moisture exchange (Michalski, 1994). Specifying such construction tolerances when placing orders for cases will help to combat many agents of deterioration.
Favour enclosed exhibit cases over open concept storage to protect sensitive displays and use pest-resistant materials and finishes to delay or inhibit penetration.
Some pesticides can act as a built-in control and repel pest activity. Cavity sprays of desiccant insecticides will reduce harbourage in otherwise hard to inspect areas. Wood treatments (such as borates) are advisable in areas susceptible to attack by wood borers.

3. Detect pests

- *Inspect:*
Inspection of the facility is carried out not only to find overt signs of pests (frass, shed skins, gnawing, etc.) but also to identify shortcomings in the facilities' ability to function as an effective pest deterrent via the means and stages. The heads of each area must be notified in advance of the inspection in order to gain access, not only to collections, but also to administration offices, and facilities such as heating plants, janitorial cupboards, etc. When conducting an inspection carry a bright

flashlight, floor plan, and notebook, and proceed methodically through all areas. It is especially worth noting holes through the building fabric, deteriorating door and window seals and flashings, unscreened vents, areas where rubbish builds up, signs of water ingress, etc. These deficiencies should be referred to the building maintenance staff and noted also as good sites for placing monitoring traps. In addition to direct examination of the facility, interview all staff, especially the sanitation, engineering, and security staff, who often go to locations that collections staff or visitors seldom frequent.

- *Monitor:*
  An effective detection system requires constant monitoring and evaluation. Mechanical traps, adhesive traps, pheromone traps, food bait traps, and light traps can be used to passively acquire knowledge on pest location, numbers, and species. Encourage all staff or visitors to report signs of pests. Determine vulnerable locations on which to focus inspections from anecdotes, the distribution of sensitive collections, or trapping studies.

- *Quarantine:*
  Agree on a quarantine policy to ensure that staff do not bypass the pest control inspection and treatment steps for incoming collections. As this policy blocks collection movement, ensure that detection and treatment can be carried out in a timely manner. Also, inform all staff of the time which treatments can require in order to allow proper scheduling of time-sensitive collections, such as loans for exhibition.

- *Record Findings:*
  Identify the pests, log the locations on a simple map, dates on which the pests were noticed, and damages caused. Obtain records left by commercial pest control operations in your institution, such as regular service contracts. These simple details can be used to construct an institutional history of infestation and track current progress. Seasonal cycles and effects of control actions can be determined from good records. Obtain information on the pests you identify to familiarise yourself with their life cycle, fecundity, food preferences, attractors, and control methods.

4. Respond

- *Sanitation:*
  Clean the infested area to reduce pest populations and remove any remaining debris.
- *Eradication methods:*
  Cleaning objects - When objects can withstand it, cleaning in hot water or dry-cleaning will eliminate insect pests. Some pests can be controlled by hand-picking, or vacuum cleaning, but the risk of missing viable stages is elevated by operator fatigue and cryptic locations of the organisms.
  Controlled Atmosphere Fumigation - Carbon dioxide (70-100%), low-oxygen (near 0.1%), and a vacuum chamber are examples of alternative fumigation methods that can eliminate insect pests from items within reasonable time (generally two weeks). These methods are attractive replacements for traditional chemical fumigants because of lower health and safety concerns and lack of toxic residuals after treatment. They have been used for nearly a century.
  Trapping - Trapping is necessary to reduce vermin populations and is best used with an aggressive exclusion program. Mechanical traps are preferred to bait stations, as corpses are removed from the building, and liability from displaced poison bait is eliminated. Pheromone trapping disrupts insect mating behaviour and can reduce populations, but is best used as an early detection or quality assurance method. The best overall trap is the delta-shaped insect trap deployed across the institution and regularly inspected.
  Thermal Control - Insects can be killed by exposure to +55°C or −20°C. While heat is quick (minutes) and cold less definitive (days) both techniques have long histories of application to a wide variety of materials (Strang, Chapter 4, this volume). Heat was used over two centuries ago, and cold for a century in the control of pests in collections.
  Insect Growth Regulators - Materials like methoprene reduce population viability by preventing complete development. However, they do hold the insect at the larval stage, which is often the most damaging stage.

Pesticide Applications - Established compounds registered for use against insects found in collections are best employed in the crack and crevice treatments, which target harbourage while preventing pesticide residues on objects.

Irradiation - Gamma irradiation has been used on a variety of collections, but is of concern to herbaria because paper is one of the materials that is easy to damage even at the low doses needed for control of insects. Further, additional handling is required to transport specimens to an off-site facility for irradiation treatment. For these reasons, irradiation is not widely used.

- *Evaluate approaches:*
Responsibilities of the person(s) charged with pest management should include the following. Collect reports, interpret data, coordinate efforts between jurisdictions to ensure the problem is adequately dealt with. Decide on response method(s) and obtain necessary equipment and training. Execute or oversee responses to ensure that the quality of work is high. Review monitoring records to assess effectiveness of control activities. Record all pest management costs, labour, losses to pests, and damages from control activity (handling, treatment, etc.) for use in planning and appropriation requests. Constantly argue to increase prevention through improving building, portable fitting, or procedural means of control. Use pest infestations to further educate staff and management about their potential to contribute to prevention.

5. Recover

Clean damaged specimens, the building, and portable fittings. This action improves the ability to detect further problems, eliminates the pests' corpses as food for other pests, and occasions staff to review the affected area once more, increasing the likelihood of detecting further problems. Repair of damages is remedial conservation treatment.

IMPLEMENTING INTEGRATED PEST MANAGEMENT

Manager Drawn from the French work "ménager" or one who does do-
mestic housework, this function has gradually been elevated to the no-
blest of levels. The strengths of the manager are continuity, stability and
the delivery of services and products from existing structures. Unfortunately managers also discourage creativity, imagination, non-linear
thinking, individualism and speaking out, an insubordinate act by
which problems are identified. (Saul, 1994)

This wry definition contains elements we experience in organizations. Germane to this topic, the root of management is housework, housekeeping, and sanitation. Improving sanitation is the
single most important element in any pest control activity. Care-
takers of stored products rank sanitation as high as 90% of pest
management activity.

Along the positive lines of Saul's definition, representatives of
all the museum functions and support staff should be involved in,
or in contact with, the institution's pest management committee
so that dangers to collections can be clearly expressed and remedial actions coordinated across jurisdictions. It is important to include these representatives so that the rationale for pest control
decisions and activities can be widely understood and procedures
agreed to within the institution. It is hard enough to control pests
without people undermining the process. Failure to include everyone leads to failure to control pests and attribution of blame for
the failure to the specific control measure, not the way in which it
was practised. Of course, not all measures are going to be completely effective, but their cumulative impact will be to reduce or
eliminate pest attack. Pest control works best when it is integrated
with the culture of preservation, not as an *ad hoc* response.

The negative aspects of Saul's definition are reflected in an organization's resistance to having to change pest control methods
from single chemical reliance even in the face of pending regulatory change. The new system has to have demonstrable efficacy in
protecting the collections, but being untried has no local credibil-
ity. Unfortunately, the existing system's efficacy is usually never
documented, so a dispassionate appraisal is hard to achieve even

ROSEWARNE
LEARNING CENTRE

when a different system is adopted. As a result, it is not uncommon to find two or more control systems applied within an institution, resulting from adherents pioneering in different directions. Contention arises as much or more from perceived risk as from measured effectiveness.

Resource Allocation

The pest management committee must, in the course of its work, periodically review its activities and judge whether they have been successful. Success is measured by the pest monitoring process. Based on this measure of experience, the committee can modify approaches and allocate resources to areas that require improvement. For pest management to succeed, ongoing monitoring of pest levels, locations, and emergence is essential. Funds for personnel and supplies for a monitoring program must be built into annual budgets. Capital improvements to effect pest control must be worked into the museum's planning.

CONCLUSION

Individual elements of Merrill's system can be compared to the steps in a modern IPM program (Table 2), highlighting steps that he did not mention. Merrill deemed his system successful and concluded "*it is not necessary to look upon the presence of from few to many of these destructive insects in any herbarium as a necessary evil and one that cannot be avoided.*"

Merrill's effective, layered protection scheme was adopted to avoid the failings of pesticide as the only control. Although naphthalene use is still strongly debated in institutions, the naphthalene/cabinet/bait system described by Merrill is worth examining. Unfortunately, when practised by others, this layered system was often simplified or allowed to degrade to sole reliance on naphthalene. Once workplace environmental legislation restricts or removes a pesticide such as naphthalene, users are left without trusted alternatives. A comprehensive integrated pest management program can continue to cope with regulatory changes, protect collections, and provide a healthier workplace.

TABLE 2. Comparison of Merrill's (1948) pest management system and the steps in a modern integrated pest management program (Strang, 1993; CCI, 1994; Rose and Hawks, 1995).

CCI Framework	Merrill's System/Strategies
MEANS OF CONTROL:	
Building	Not mentioned
Portable fittings	Tight cabinets
	Fumigation equipment
Procedures	Curatorial commitment and responsibility, routine inspection of traps and fumigation. Notification and cautioning of loan recipients about the untreated nature of the loaned specimens and expectation for their care.
STAGES OF CONTROL:	
Avoid	Not mentioned
Block	Tightly sealed metal cabinets. Prophylactic use of naphthalene as a repellant. Bundles of attractive plants to distract and delay infiltrating pests.
Detect	Minimum quarterly inspections of plant bundles/traps.
Respond	Fumigation of incoming loans and acquisitions. Quart erly fumigation of plant bundles/traps. Heat and vacuum alternatives to fumigation.
Recover	Not mentioned

ACKNOWLEDGMENTS

The author would like to thank D. Metsger, J. Waddington, J. Bull, S. Byers, and S. Elliot for their comments during the preparation of this paper, which drew heavily on the author's perpetual draft technical bulletin.

LITERATURE CITED

[CCI] CANADIAN CONSERVATION INSTITUTE. 1994. Framework for Preservation of Museum Collections. Canadian Heritage, Government of Canada, wall chart.

Ertter, B. 1999. Elements of herbarium layout and design. Pp. 119-145 in D. A. Metsger and S. C. Byers, editors. *Managing the Modern Herbarium*. Society for the Preservation of Natural History Collections, Washington, DC, xxii+384 pp.

Lull, W. P. and B. P. Moore. 1999. Herbarium building design and environmental systems. Pp. 105-118 in D. A. Metsger and S. C. Byers, editors. *Managing the Modern Herbarium*. Society for the Preservation of Natural History Collections, Washington, DC, xxii+384 pp.

Merrill, E. D. 1948. On the control of destructive insects in the herbarium. Journal of the Arnold Arboretum 29:103-110.

Michalski, S. 1994. Leakage prediction for buildings, cases, bags, and bottles. Studies in Conservation 39:169-186.

[NTP] National Toxicology Program. 1992. *Toxicology and carcinogenesis studies of naphthalene in B6C3F$_1$ mice*. US Department of Health and Human Services, National Institutes of Health Publication 92-3141, 198 pp.

Rose, C. L. and C. A. Hawks. 1995. A preventive conservation approach to the storage of collections. Appendix II: Selections from the wall chart: framework for preservation of museum collections. Pp. 15-17 in C. L. Rose, C. A. Hawks, and H. H. Genoways, editors. *Storage of Natural History Collections: A Preventive Conservation Approach*. Society for the Preservation of Natural History Collections, Iowa City, IA, 448 pp.

Saul, J. R. 1994. *The Doubters Companion, A Dictionary of Aggressive Common Sense*. Viking, Toronto, 342 pp.

Scott, E. W., W. S. Abbott, and J. E. Dudley, Jr. 1918. *Results of experiments with miscellaneous substances against bedbugs, cockroaches, clothes moths, and carpet beetles*. US Department of Agriculture, Bulletin No. 707, 36 pp.

Strang, T. J. K. 1993. *Museum pest management*. Draft Technical Bulletin, Canadian Conservation Institute, 31 pp.

Strang, T. J. K. 1998. Sensitivity of seeds in herbarium collections to storage conditions, and implications for thermal insect pest control methods. Pp. 81-102 in D. A. Metsger and S. C. Byers, editors. *Managing the Modern Herbarium*. Society for the Preservation of Natural History Collections, Washington, DC, xxii+384 pp.

Williams, S. L. and S. B. McLaren. 1990. Modification of storage design to mitigate insect problems. Collection Forum 6(1):27-32.

APPENDIX I

Additional Resources

The following books give information on applied pest control, references into the core literature on pest species, and assist in the identification of unfamiliar pests. These volumes are very comprehensive and provide a convenient summary of the necessary details about the pests which affect collections. Imholt's book is valuable when extending pest control considerations into building level design and detailing, a practice which is increasing within the collections community as collections are re-housed.

EBELING, W. 1978. *Urban Entomology.* Revised edition. University of California, Berkeley, 695 pp.

GORHAM, J. R., editor. 1991. *Ecology and Management of Food-Industry Pests.* FDA Technical Bulletin 4. Association of Official Analytical Chemists, Arlington, Virginia, 595 pp.

GORHAM, J. R., editor. 1991. *Insect and Mite Pests in Food, An Illustrated Key.* 2 Volumes. US Department of Agriculture and US Department of Health and Human Services, Washington, 767 pp.

IMHOLT, T. J. 1984. *Engineering for Food Safety and Sanitation. A Guide to the Sanitary Design of Food Plants and Food Plant Equipment.* Technical Institute of Food Safety, Crystal, Minnesota, 283 pp.

MALLIS, A. 1997. *Handbook of Pest Control. The Behavior, Life History and Control of Household Pests.* Eighth Edition. Mallis Handbook and Technical Training Company, 1456 pp.

The following books and articles focus on cultural collections. While smaller in scope than the predominantly agricultural focus of the previously listed books, these volumes limit the number of pests under consideration, yet cope better with the particular considerations of cultural collections.

ÅKERLUND M. 1991. *Ångrar - finns dom...? Om skadeinsekter i museer och magasin.* Svenska museiföreningen, 207 pp.

BROKERHOF, A. W. 1989. *Control of Fungi and Insects in Objects and Collections of Cultural Value.* Amsterdam, Central Research Laboratory, 77 pp.

JESSOP, W. C. 1995. Pest Management. Pp 211-220 in C. L. Rose, C. A. Hawks and H. H. Genoways, editors. *Storage of Natural History Collections: A Preventive Conservation Approach*. Society for the Preservation of Natural History Collections, Iowa City, IA, 448 p.

PINNIGER, D. 1994. *Insect Pests in Museums*. Third edition. London, Archetype, 58 pp.

RUST, M. K. AND J. M. KENNEDY. 1993. *The Feasibility of Using Modified Atmospheres to Control Insect Pests in Museums*. J. Paul Getty Trust, 125 pp.

STORY, K. O. 1988. *Approaches to Pest Management in Museums*. Smithsonian Institution, Washington, DC, 165 pp.

WELLHEISER, J. G. 1992. *Nonchemical Treatment Processes for Disinfestation of Insects and Fungi in Library Collections*. IFLA Publications 60. K.G. Saur, Munich, 118 pp.

ZYCHERMAN, L. A. AND J. R. SCHROCK. 1988. *A Guide to Museum Pest Control*. Association of Systematics Collections, Washington, DC, 205 pp.

CHAPTER 4

Sensitivity of Seeds in Herbarium Collections to Storage Conditions, and Implications for Thermal Insect Pest Control Methods

TOM J.K. STRANG

Abstract.—Throughout this century, herbarium staff have increasingly used convection heating, refrigeration, and microwaving for the control of insect pests. Emphasis within the resulting literature has been placed either on the ability of the methods to control pests, or on concerns about the effect of the treatment on herbarium materials. In order to critique thermal pest control methods, damage thresholds for collections have to be established. To compensate for the lack of defined damage thresholds in the botanical literature, mortality data for seeds of dried vascular plant specimens, drawn largely from the agricultural literature, are used here as an early indicator of material damage to herbarium specimens in general. Since thermal treatments are merely an extension of the storage environment, out of the same data it is possible to extract ideal storage conditions that are based on measured performance. Given the potential for loss of seed viability due to normal herbarium practices and natural longevity, the additional risks to seed viability posed by thermal pest control methods are considered to be minor compared to the certain ravages by insect pests.

INTRODUCTION

Herbarium staff, as with all museum professionals, are ethically bound to consider the effects of any pest control methods on collections so that they do not reduce the collection's research potential in efforts to save it from a highly destructive agent such as insects. Conversely, potential for losses to parts of the collection must be balanced against the overall good gained from a treatment that protects the entire collection. This paper puts this ethical requirement for thermal pest control in perspective with the available knowledge on seed viability. Herbaria have the interesting twist that they are likely to be the only living collection in many natural history museums because of the seeds or spores attached to, or cultures associated with, the specimens. A seed's complex chemistry makes it more broadly susceptible to damaging agents than any one macromolecule, and because the seeds may be viable, one can measure changes in their viability as a threshold indicator of potential deterioration and damage to other tissues.

Botanical collections have, in the past, been sampled for viable seed (Harrington, 1972). With vascular plant collections, this allows researchers to entertain the prospect of germinating some seeds from the collection to perpetuate a species, or to extend botanical knowledge (Hill, 1983). Womersley (1981) states that germination from seed is the best approach to the study of seedlings and floral development. With the exception of fungal collections, herbaria are not primary sources of living material. They are, however, a potentially valuable source of material for macromolecular studies, and so pest control techniques must certainly respect this potential use of the collections.

Within the botanical literature, seed death and physical damage have been cited as an inhibiting caveat to the use of some methods of thermal pest control (Hill, 1983; Forman and Bridson, 1989; Stansfield, 1989; Egenberg and Moe, 1991; Gunn, 1994). These assertions seem to contradict those to be found in other disciplines. The agricultural literature, for instance, contains many references to conditions affecting seed viability that can be applied to the herbarium context.

By treating seeds as dosimeters that log cumulative damage to the collections, this paper will seek to determine if any general, mixed group of herbarium specimens will survive either high or low temperature control aimed at eliminating insects, and if thermal techniques are marginalized by an overwhelming adverse effect on the treated specimens.

In considering the adverse effect on the specimens I will distinguish between deterioration and damage. Deterioration is often described chemically (e.g., loss of degree of polymerization in organic molecules) and is incremental and unavoidable although we may modify its rate. Damage represents actual harm to specimen integrity, and so utility. The description of damage is specimen-dependant; however, for purposes of this discussion, it refers to the loss of taxonomically important characters. The scale of damage to botanical material will range from the loss of macroscopic features to the loss of intact DNA. Examining seed mortality will show if this range of damage is occurring, since damage at macroscopic, microscopic, or molecular levels will surely contribute to seed senescence and death.

HISTORICAL BACKGROUND

Thermal treatments, among the so-called new or alternative methods in use today, have actually been used successfully for a longer time than the more familiar chemicals found on or around herbarium specimens: mercuric chloride, naphthalene, paradichlorobenzene, lauryl pentachlorophenate, etc. Table 1 gives a very brief history of thermal pest control applications to natural history collections. In the first four decades of the twentieth century, before the advent of modern fumigants and pesticides, heating greenhouses and entire grain mills was recognized by millers and their insurers, and by some botanists, as the most efficacious and cost effective method (Dean, 1911; Back, 1920; Cressman, 1935; O'Neill, 1938) . There are mills that still operate using heat disinfestation. The argument restricting renewed expansion of this scheme revolves not around efficacy, but around the required capital investment, as compared to the continuing low immediate cost of fumigant use.

Attention has returned to the use of thermal pest control methods in collections because many favoured pesticides and fumigants

Table 1. Brief summary of publications referring to thermal control of insects in natural history collections.

AUTHOR	YEAR	COMMENTS
Kuckahn	1771	recommended heating bird specimens to kill insects
Howard	1896	described disinfesting fur with −8°C storage
Dean	1911	expanded on other work from 1883 and reported no effect on germination of grain exposed in a mill heated to 66°C over eight hours
Bovingdon	1933	killed *Lasioderma serricorne* in cigars and bales of tobacco with −6°C
Cressman	1935	used gas burners and electric fans to disinfest a library
O'Neill	1938	wrote on the outfitting of herbarium cabinets to heat treat their contents to 60–70°C within four to five hours
Crisafulli	1980	recommended −18°C to kill *L. serricorne* in herbarium collections
Watling	1989	recommended 42°C heat to kill mycophagous insects before accessioning the fungi; also warned of −18°C having failed to kill four species of insect
Miller and Rajer	1994	used low winter temperatures in 1985 to chill a herbarium below −18°C

have been restricted by regulatory changes and thus removed from the hands of collections staff. Thermal methods are attractive because they involve no toxins or toxic residues, are largely available to anyone, and need not depend on an expensive or governmentally regulated infrastructure. By applying the principles of thermal control, large quantities of material can be preserved at low unit cost and on short notice. Strang (1992) provides a comprehensive treatment of the efficacy and history of thermal insect pest control.

Within the botanical literature, articles on microwave heating provide the most discussion on a thermal treatment in the herbarium context so they will be reviewed here. The effect of microwaves on insects has been documented since the 1920s and work with them performed by agriculturalists since the 1940s (Nelson and Stetson, 1974). This work concentrated on finding whether there was an appropriate microwave frequency to affect pests which minimised damage from heating to the surrounding commodity, and on trying to determine whether efficacy was due to more than general heating. At 2450 MHz frequencies there was no differential heating; both insect and commodity needed to be heated above 60°C to kill the pests, a condition identical to convection heating. At lower frequencies (1 to 50 MHz), microwaves did create the differential heating effect for some insects and commodities, but it was not universal. While some researchers had noted the ability of microwaves to kill pests without affecting grain germination, provided the moisture contents were low, general adoption of microwaves was prevented by energy costs that exceeded the cost of conventional fumigation by a factor of five (Nelson, 1967). When microwaves became readily available during the 1980s in the form of rapid cooking devices (2450 MHz), botanists and other collection professionals naturally tried these machines for pest control and drying specimens.

Hall (1981) advocated microwaves as a fast method to kill insects which alleviated pressures created by the need to quickly associate suspect material with clean holdings. No adverse effects were mentioned, so insect damage to the entire collection obviously took priority over any particular loss to the dry specimen, and the practice was used for a year prior to Hall's publication. Hardin (1981) treated all accessions with microwaves. With a microwave unit right in the collection, he also treated the entire collection annually, taking one hour per cabinet.

Clearly there have been situations in which botanists have desired the dramatic effects of microwaves on a succulent specimen. "The material is 'done' when it has a flaccid, water-soaked appearance. It can then be arranged on a newspaper and pressed in the usual way..." (Fuller and Barbe, 1981). However, Baker et al. (1985) described the damage threshold: "discoloration and singeing of the spines", and sticking to absorbent towels, which they also tell how to avoid.

Other microwave users started noting their concerns by indicating damage thresholds for their collections from microwave treatments: Hill (1983) demonstrated damaging loss of seed viability on old specimens. Philbrick (1984) echos the seed concern, and postulates macromolecule and fine morphological deterioration. Bacci et al. (1985) demonstrated that morphology visible in a light microscope and specimen colour were preserved adequately after microwaving, but ultrastructure was adversely affected compared to room temperature drying. Jacquin-Dubreuil et al. (1989) demonstrated that relevant morphology and alkaloids were preserved through microwave drying. Arens and Traverse (1989) demonstrated no additional loss of pollen or spores by microwaving after standard drying practice, but much higher losses in fresh material to microwaving when compared to air drying. Pyle and Adams (1989) found damage to grass material after three minutes of microwaving, but did not try dry samples, or 60°C convection heat. Diprose et al. (1984) noted that germination actually increased in some seed that was heated by microwaves but cited losses in other tests.

In the end, botanists working on herbarium collections independently rediscovered what the agriculturalists had reported in previous decades about microwave treatment: microwave heating is far less controllable than convection heating and so it is easy to overheat and damage tissues.

I am concerned about the differences between the conclusions drawn in herbarium literature and those found in agricultural literature concerning the effects on seeds of convection heating and cooling (often called freezing) used for the control of insect pests. Because herbarium staff have raised seed viability as a reason for not using thermal methods, I will devote much of this paper to focusing on factors affecting seed viability.

EFFECT OF HERBARIUM PRACTICES ON SEED VIABILITY

Because thermal treatments are merely an extension of the storage environment, it is first necessary to consider whether herbarium practices damage seed viability, our allegorical figure for a collection's physical survival.

Developmental Stage of the Seed When Collected

It must be recognised that within a herbarium collection, the plants may not have been collected at the correct time to ensure seed viability. If they were already dry on the plant, seeds may

FIGURE 1. Damage thresholds for seeds in herbarium storage. Equilibrium moisture content (EMC) is mapped to corresponding relative humidity by sorption isotherms. Sorption isotherms from Harrington (1972), Blum (1973), and Iglesias and Cherife (1982). Seed damage thresholds from Howe (1965), Harrington (1972), and Priestley (1986). Paper threshold from Brandon (1980). Humidification system threshold from Harriman (1990).

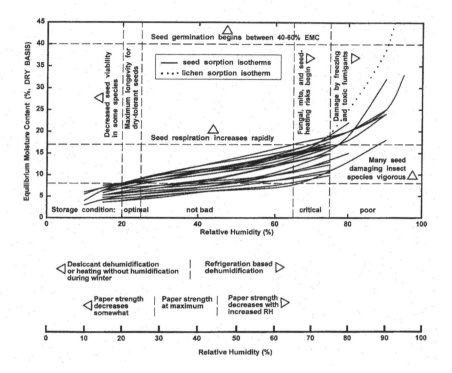

store reasonably well in herbarium conditions. Seeds that are wet (e.g., found in fruit or not quite mature) must dry to equilibrium below 65% relative humidity (RH), otherwise they will succumb to fungal attack and respiration heating (Fig. 1). After fertilization, seeds pass through several named stages: zygote, embryo (often in liquid endosperm), ovule (early, late), seed (immature, ripening, mature, senescent, dead). The following benchmarks of survival are generalized in Harrington's review (1972). The late ovule stage is the earliest stage at which seeds normally survive when separated from the parent plant. Immature seeds may germinate on sowing when they are fresh but will die if they are dried out. The immature stage has a maximum storage life rated in hours or a few days. Maturity, for some seeds, occurs when they reach their maximum dry weight. Other seeds mature their embryos or lose dormancy during a later drying phase. Most mature seeds can be dried and stored.

Dryness in Storage

Whether a seed is on the plant or stored separately, its longevity (and demonstrably the longevity of all organic matter) is related to moisture content. Moisture content is mapped to corresponding RH by sorption isotherms (Fig. 1). Strang (1995) outlines the relevance of sorption isotherms to thermal treatments on common collection materials within tight enclosures. Appropriate humidity norms for general museum collections and the underlying technical reasons are discussed in Michalski (1993).

For optimal, long-term storage in moisture proof enclosures, seed packagers strive for a dry-weight equilibrium moisture content (EMC) that matches the 20–25% RH range (Harrington, 1972). The measurement of moisture content in seeds has traditionally been on a wet basis, whereas many scientific disciplines routinely use dry basis. For this paper, moisture contents have been translated to dry basis, although the differences at low EMCs are fairly small.

Below 20% RH, Harrington (1972) described the auto-oxidative phenomenon to be mildly deleterious. Priestley (1986) proposes an alternate mechanism for lower germination around imbibitional stresses with examples of seed unaffected by extreme dryness. Whatever the cause, there are measured losses in viability at RHs below 20% for some seeds.

Above 25% RH seed deteriorates faster from chemical processes, although the 25% to 65% RH range is not overly detrimental to viable dried seed held over the short term (Owen, 1956). Above 65% RH, the probability of mould increases: three months near 70% RH, two weeks near 80% RH, and a couple of days at over 90% RH (Michalski, 1993).

To an extent, dry storage of seed restricts insect activity (Fig. 1). While there are a few seed-destroying insects that still thrive on very dry seed (e.g., *Trogoderma granarium* Herts, *Callosobruchus* spp.), many species become moribund on dry seed, that is seed at equilibrium with less than 30% RH (Howe, 1965).

Of course, herbarium specimens are more than just seeds, and low RH does not necessarily protect the specimens from all pests, although it is now often suggested as a means of control (Lull and Moore, this volume). Bridson and Forman (1992) mention environmental conditions for herbarium storage, but only as a contributory element to pest control and "staff efficiency". Unfortunately, their suggested 20° to 23°C, 40 to 60% RH regime is not likely to greatly affect pests (see Howe, 1965; and Fig. 1).

At this point, I have to depart from the narrow seed allegory and discuss the connections to wider issues in the context of humidity norms for herbarium storage. Since herbaria use paper mounting sheets and envelopes, a parallel issue is paper strength or brittleness. Brandon (1980) summarized strength losses in burst and tensile tests performed below 30% and above 50% RH (Fig. 1, lower graph). Fold endurance rises with humidity until 70% to 90% RH. Maintaining humidity below 30% would not contribute to preserving paper from the stress of improper handling.

Reducing RH does slow chemical deterioration (measured as strength loss) and paper lifetime is doubled by a drop in storage RH from 50% to 30% (Michalski, 1993). This is in contrast to Bridson and Forman's (1992) ideal conditions of "20–23°C at ± [sic] 55% humidity." If one wishes to extend the material lifetime of a collection, specifying as low an RH as is feasible appears to be the justifiable and quantifiable specification, not artificially holding 55% RH.

How does changing humidity affect the assemblies of plant and paper we call mounted specimens? To get a sense of the effect, while moisture content of seed increases with increased RH of the seed store, fluctuations in seed moisture content are moderated by the seed coat, surrounding paper media such as envelopes and sheets, and the closed cases. The response time of different species' seeds exposed to altered humidity range from one day to eight weeks at room temperature (Owen, 1956). The measured time to equilibrium doubled with a 10°C drop in temperature (Whitehead and Gastler, 1946–7). The effect of any surrounding storage materials will be to further slow any shift in equilibrium. This is important to remember if one is concerned about a short duration swing in humidity, either in the herbarium environment, or during a thermal or controlled atmosphere pest eradication treatment. For a larger treatment of moderating humidity effects during thermal treatment, see Strang (1995).

Egenberg and Moe (1991) demonstrate damage to adhesive-dot mounted *Primula* sp. cycled between 95% and 15% RH. These conditions are called normal by the authors. However, in contrast to their conclusions, this range of humidity cycling does not occur with specimens treated for pests at either low or high temperatures (as defined in Strang, 1992) when the specimens are bagged at median relative humidities (Strang, 1995). Michalski (1993) demonstrates why less than 25% RH and greater than 75% RH are better termed extremes when discussing the humidity response of organic materials.

To obtain building-wide, room, or case humidity below 40% RH at human comfort temperature, one is usually required to use desiccant dehumidification, while control above 40% RH is possible using more conventional refrigerant systems (Fig. 1; Harriman, 1990). Note that in temperate climates one can achieve low RHs in winter months by heating without humidification. Humidification of the collection space much higher than 30% RH during winter heating months is not really justified for botanical collection preservation.

For maximum extension of specimen longevity when humidities are uncontrollably high, one should apply local solutions such as placing dried specimens with desiccants in sealed enclosures. Sufficiently dried material in a tight enclosure can remain unaffected by external high humidities for years (Michalski, 1994; Strang, 1995).

Drying Temperature and Method

While terminal moisture content of seed is the primary factor in prolonging viability, drying methods will also affect the result. Air drying in the sun can expose objects to temperatures 20°C to 40°C higher than ambient (Yamasaki and Blaga, 1976). Absorption of infrared radiation, especially by dark objects, and the insulating ability of stagnant air easily creates temperatures of 50°C to 60°C. These principles were used to advantage in the design of a solar plant drier (Sinnot, 1983), and for eliminating textile pests (Strang, 1995). Mechanically-assisted seed drying with fans relies on the ambient humidity being lower than that over the seed. Under damp conditions, seed may not dry fast enough to protect it from losses caused by fungal attack or partial germination.

Safe maximum temperatures for drying newly harvested crop seeds with heated air are 45°C for grain, beets, and grasses, and 35°C for vegetable seed (Harrington, 1972). These temperatures match those selected for use in herbarium drying cabinets to reduce specimen browning. Drying too fast leads to cracking of the seed coat as it shrinks over the damper core of the seed, but prolonged heating at the higher EMC also damages viability (Harrington, 1972). Dry seed can withstand higher temperatures. Figure 2 plots the relationship between EMC and maximum safe drying temperature from data in reviews by Harrington (1972) and Owen (1956). A boundary is created by the intersection of low and high percentage germination data. Not surprisingly, commercial recommendations delineate the boundary for safe exposure at higher moisture contents.

Very dry seed survives temperatures near 100°C for short periods. This fact is exploited by systems used in grain mills that kill insect pests by passing seed through heated enclosures. Restricting the length of thermal exposure is not as critical as maintaining a low moisture content, even in short-lived seed (Rocha, 1959). Standard herbarium practice for drying damp plants and their attached seeds uses fan-blown air at 45°C. While this may be too high for some fresh seeds, it lies within the commercial norm for others (Fig. 2). It appears that elevating temperatures of herbarium specimens to 55⁻60°C to kill insects (Strang, 1992) would not pose a significant risk to germination of dried viable seed. It might be wise to limit thermal treatments to specimens that are acclimated to 50% RH or less at room temperature, which would allow a margin of safety. Heat treating damp specimens to kill insects should be carried out only after first drying them down. However, if the specimens have been held for a prolonged time in elevated humidity (such as herbaria in the tropics), the damage

Figure 2. Heat damage to seed. Drying temperatures plotted against equilibrium moisture content (EMC). Data from Owen (1956) and Harrington (1972). Relative humidities at 20% and 65% are plotted vertically from their equivalent EMC (based on sorption isotherms in Fig. 1) to enclose a zone that represents optimal to fair herbarium storage (marked B). Overly dry storage is marked A, and post-harvest/in vivo moisture levels are marked C. A damage threshold is created by the intersection of low and high percentage germination data, which continues into commercial seed drying specifications at higher moisture contents.

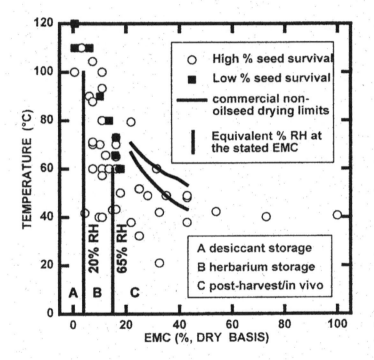

from moisture-mediated deterioration may well have occurred and this restriction will prove unnecessary. Since the RH over the seed and other organic matter in a vapour barrier enclosure increases as the temperature rises, treatment of specimens that are acclimated to 65% RH at room temperature might at first glance be at risk of a mould outbreak. This risk is actually small. These samples might reach 80% RH during heating, but the time for mould response at 80% RH is long (10 days), and the treatments are very short (a few hours). Also, at 55‑60°C, many of the moulds, with the exception of thermophiles, are unable to function

(Michalski, 1993). See Strang (1995) for a longer discussion on mould and condensation risk during thermal treatments and how to avoid it.

As a historical note, in addition to drying, brief immersion of seed in water near 55°C was a well-developed technique to eliminate fungal damage prior to planting with little change in germination, as were hot water sprays on living plants (Bourcart, 1926).

Low Storage Temperatures

Seeds, like other organic tissues, have a storage lifetime that doubles roughly with a 5°C drop in temperature. Harrington (1972) points out that this rate is known for seeds tested between 0°C to 50°C but there is little data outside these limits. Figure 3 shows the effect of storage temperature on seed viability. These experiments may be carried out either in a refrigerated container at

FIGURE 3. Effect of storage temperature on viability of seeds stored at the same equilibrium moisture content. Each line represents seeds of one species. Data from Owen (1956).

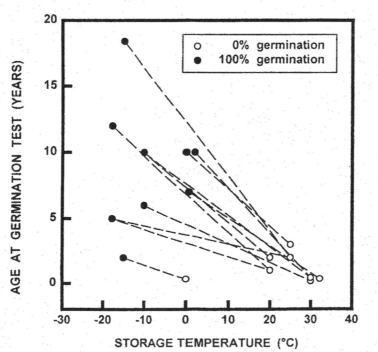

constant relative humidity, or hermetically sealed and then chilled. In the latter case, the seed moisture content rises a little and equilibrium RH drops.

Seed that is dried to less than 14% EMC can be safely stored at temperatures below freezing. Damage (loss of viability) is only slight near 20% EMC (Harrington, 1972). 20% EMC is roughly the fibre saturation point of many wood xylem cells. The often-raised issue of water freezing and damaging organic matter during low temperature treatment receives no support from nuclear magnetic resonance studies which show no freezing activity between 0°C and –30°C in organic matter with EMCs less than 25% (Pichel, 1965; Toledo et al., 1968; Nanassy, 1978).

By the mid 1900s, researchers concluded that while there was significant improvement of viability in seeds that tolerate drying when stored between 20% and 30% RH, and at temperatures near 0°C, the lowering of either temperature or humidity was good enough for agronomic purposes (Owen, 1956). This pragmatic choice reduced the need to use temperatures much below 0°C. After all, why look for something better than good enough? Where commercial pressures are not as relevant, temperatures near –15°C to –20°C are used to preserve seed, and liquid nitrogen (–196°C) is also employed (Priestley, 1986).

Pyle and Adams (1989) recovered comparable DNA fragments from fresh plant material and samples stored for five months at –20°C, 4°C, and 42°C. Low temperature pest control is not likely to affect DNA recovery.

Natural Seed Longevity

The published longevity of seeds stored indoors are plotted in Figure 4. Long-lived seeds, retaining viability for more than 10 years at ambient conditions, are often characterised by dryness at maturity (under 4% EMC) and a hard seed coat that requires abrasion prior to germination. The range in viability of these seeds spans the time that many herbarium collections have existed, from decades to centuries — longer in the case of some dated archaeological recoveries.

Even if collected at an appropriate stage, some seeds cannot be saved for more than a short time. Harrington (1972) lists this varia-

FIGURE 4. Percent germination over time of a mixed collection of 584 species of seed stored indoors in botanical or laboratory collections. Data from Owen (1956), Harrington (1972), and Priestley (1986).

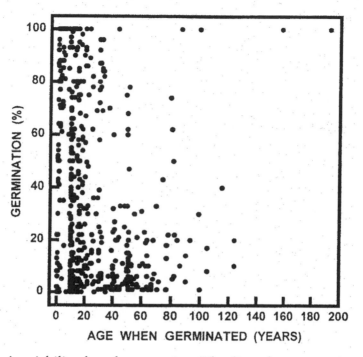

tion in viability by plant species. The list of poor survivors is dominated by aquatics, nut trees, and tropical species. Seed senescence in these species occurs between 2 weeks and 5 years. Other seeds are very short lived as they do not tolerate any drying, even once mature.

Factors that contribute to lower germination of seeds include stresses on the mother plant (Harrington, 1972); size less than 1 mm, and study field conditions (Priestley, 1986); and unknown temperature optima ranging from 0.5°C to 50°C (Mayer and Poljakoff-Mayber, 1963).

Of the 584 species stored indoors and tabulated by Harrington (1972), 128 survived less than five years. Only five species are listed as being damaged by temperatures under 0–10°C. Sixty-two species are listed as being intolerant to drying, i.e., they are viviparous. From this alone, one must expect less than three-quarters of species in a mixed collection to contain viable seed. The germination trend in Figure 4 is not encouraging for old collections, but establishes a reasonable probability that some of the holdings in herbaria are yet viable. Therefore, botanists questioning potential effects of treatments on seed viability have some reason to be concerned.

Impact of Fumigants on Seed Viability

Controlled atmospheres are used to kill insect pests by creating either low levels of oxygen (0.1%) or high concentrations of CO_2 (>60%) (Annis, 1986). Both of these conditions contribute to seed dormancy (Harrington, 1972). Experiments on seed storage in these cover gasses showed them to be beneficial to neutral in extending seed life (Owen, 1956). Either gas must be applied for a minimum of three weeks at room temperatures (20–30°C) to ensure efficacy against seed boring insects (Annis, 1986). Efficacy is rather temperature sensitive, and higher temperatures are preferred.

Somatic cell abberations have been noted in germinated seed that had been stored at room temperature in air, carbon dioxide or nitrogen, or heat treated along the boundary area shown in Figure 2 (Owen, 1956). Chromosome errors increase with seed age, but are reduced with cool, dry storage. These changes were not considered to be serious for commercial ventures, but are possibly of interest to more specific investigations (Harrington, 1972). These errors will have occurred in aging specimens under normal herbarium conditions.

Toxic gas fumigants are not used on high moisture content seed as the seed is more likely to succumb to the fumigant (Fig. 1). A dry herbarium collection would not be as susceptible. Naphthalene has reduced germination of some seeds but not others over trials of one year, and herbicide vapours decreased seed germination (Owen, 1956).

CONCLUSIONS ON THE RISKS TO SEEDS FROM THERMAL CONTROL TREATMENTS

The vulnerabilities of seeds cited in this paper demonstrate that seeds susceptible to damage from thermal pest control methods are likely to have succumbed to standard herbarium processing and storage anyway. With the exception of treatment in micro-waves, most viable dried seeds will survive the necessary pest control temperatures under controlled application. Microwave heating is far less controllable than convection heating; overheating is easy and so microwaving is often fatal to seed. Convection heating without tight containment of the sample to maintain moisture certainly desiccates material, causing damage. Likewise, unprotected samples in freezers face an unnecessary mould risk in the event of mechanical failure. The only published description of damage to a specimen on a herbarium sheet (Egenberg and Moe, 1991) is actually limited to extreme humidity change, not temperature. Such humidity change does not occur in thermal treatments of bagged material. *Thermal treatments should always be carried out in vapour barrier enclosures.*

Forman and Bridson (1989) make no mention of convection heating, yet it clearly avoids the majority of problems associated with microwaving. Convection heating has been used longer than any other thermal method to prepare herbarium specimens for storage. Convection heating is incapable of elevating specimens above the controlled-supply air temperature, and readily disinfests blocks of specimens for use by researchers in an hour (Strang and Shchepanek, 1995). A few sheets are done in minutes.

The following observations are offered to assist those debating the use of thermal pest control in herbarium collections:

• Practical experience shows no damage to herbarium collections in moisture barrier enclosures when low temperature is used for pest control, and chemical deterioration is actually slowed for the duration of the treatment. Refrigeration, including temperatures below 0°C, usually prolongs seed viability, especially shorter-lived varieties (Fig. 3). Low temperature control of insect pests is possible on all seeds that are tolerant to drying. Researchers looking at macromolecules (including enzymes and DNA) generally prefer low temperature control to elevated temperature; they routinely use low temperatures to store their materials.

• Available data indicate that properly dried viable seeds can withstand the convection oven temperatures used to kill insect pests (Fig. 2). Treatment times can be very short if the material is properly organized (Strang and Shchepanek, 1995). The material undergoes no significant change in dimension or moisture content if tightly bagged (Strang, 1995). Counter to Stansfield's (1989) criticism, proper technique avoids damage to objects and specimens during heat treatments.

The loss of specimen longevity due to increased oxidation of cellulosics during a one hour heat treatment at 60°C every 10 years in perpetuity is 0.15% (S. Michalski, CCI, pers. comm.). Thus, heat treatments can be used sporadically when the need arises without too great a concern for reducing specimen longevity. More precisely, the rate of loss due to pests (if measured) can be compared to the projected loss due to the pest control treatment.

- While some authors have demonstrated benefits, microwave heating is not favoured as a pest control method as it is harder to control the peak temperature, metallic inclusions introduce fire hazards, and chamber sizes are often too small to accommodate standard herbarium sheets.

- Properly dried viable seed will likely withstand the low and high temperatures necessary to ensure insect mortality. For seeds that withstand drying, humidity is a dominant factor in seed longevity. If preservation of seed is an objective, herbarium staff should look to their environmental norms to maximize seed longevity and insure the specimen is within these norms prior to thermal treatment. Drier storage conditions also increase the lifetime of plant and mount materials, and slow some insect pests (Fig. 1). Those seeds that do not tolerate drying will already be dead in herbaria unless special procedures were taken to preserve viability at the time of collection.

- Eventual species-specific seed senescence (Fig. 4) suggests that if herbarium specimens are to be used as a seed source, this should be done sooner rather than later.

In conclusion, the risks of using thermal insect pest control methods for herbarium samples are minor compared to the certain ravages by insect pests.

ACKNOWLEDGMENTS

I thank M. Shchepanek and M. Bouchard of the Canadian Museum of Nature for their assistance, and R. Rabeler, T. Dickinson, J. Waddington, and D. Metsger for their critical reviews of this paper.

LITERATURE CITED

ANNIS, P. C. 1986. Towards rational controlled atmosphere dosage schedules: a review of current knowledge. Pp. 128–148 in E. Donahaye and S. Navarro, editors. *Proceedings of the 4th International Working Conference on Stored-Product Protection*. Tel Aviv, Israel, xii+668 pp.

ARENS, N. C. AND A. TRAVERSE. 1989. The effect of microwave oven-drying on the integrity of spore and pollen exines in herbarium specimens. Taxon 38:394–403.

BACCI, M., A. CHECCUCCI, G. CHECCUCCI, AND M. R. PALANDRI. 1983. Microwave drying of herbarium specimens. Taxon 34:649–653.

BACK, E. A. 1920. *Insect control in flour mills*. USDA Bulletin 872, Washington, DC.

BAKER, M., M. W. MOHLENBROCK, AND D. J. PINKAVA. 1985. A comparison of two new methods of preparing cacti and other stem succulents for standard herbarium mounting. Taxon 34(1):118–121.

BLUM, O. B. 1973. Water Relations. Pp. 381–400 in V. Ahmadjian and M. E. Hale, editors. *The Lichens*. Academic Press, New York.

BOURCART, E. 1926. *Insecticides, fungicides and weed killers*. Trans. Second edition, revised by T. R. Burton. Ernest Benn Limited, London.

BOVINGDON, H. H. S. 1933. Report on the Infestation of Cured Tobacco in London by the Cacao Moth *Ephestia elutella* Hb. Empire Marketing Board Report No. 67.

BRANDON, C. E. 1980. Properties of paper. Pp. 1715–1972 in J. P. Casey, editor. *Pulp and paper, chemistry and chemical technology*. Vol. 3. Third edition. John Wiley and Sons, New York.

BRIDSON, D. AND L. FORMAN. 1992. *The Herbarium Handbook*. Revised edition. Royal Botanic Gardens, Kew.

CRESSMAN, A. W. 1935. Control of an infestation of the cigarette beetle in a library by the use of heat. *Journal of Economic Entomology* 26:294–295.

CRISAFULLI, S. 1980. Herbarium insect control with a freezer. Brittonia 32:224.

DEAN, G.A. 1911. Heat as a means of controlling mill insects. Journal of Economic Entomology 4:142–161.

DIPROSE, M. F., F. A. BENSON, AND A. J. WILLIS. 1984. The effect of externally applied electrostatic fields, microwave radiation and electric currents on plants and other organisms, with special reference to weed control. The Botanical Review 50:171–223.

EGENBERG, I. M. AND D. MOE. 1991. A "stop-press" announcement. Damage caused by a widely used herbarium mounting technique. Taxon 40:601–604.

FORMAN, L. AND D. BRIDSON. 1989. *The Herbarium Handbook*. Royal Botanic Gardens, Kew.

FULLER, T. C. AND C. D. BARBE. 1981. A microwave-oven method for drying succulent plant specimens. Taxon 30:867.

GUNN, A. 1994. Past and current practice: the Botanist's view. Pp. 11–14 in R. E. Child, editor. *Conservation and the Herbarium*. The Institute of Paper Conservation, Leigh Lodge, Worcestershire.

HALL, D. W. 1981. Microwave: a method to control herbarium insects. Taxon 30:818–819.

HARDIN, J. W. 1981. Pest control with a microwave oven. Herbarium News 1(2):1–2.

HARRIMAN, L. G. III. 1990. *The Dehumidification Handbook*. Second edition. Munters Cargocaire, Amesbury, MA.

HARRINGTON, J. F. 1972. Seed storage and longevity. Pp. 145–245 in T. T. Kozolowski, editor. *Seed Biology*. Vol. 3. Academic Press, New York.

HILL, S. R. 1983. Microwave and the herbarium specimen: potential dangers. Taxon 32:614–615.

HOWARD, L. O. 1896. Some Temperature Effects on Household Insects. USDA Division of Entomology, Bulletin 6, New Series:13–17.

HOWE, R. W. 1965. A summary of estimates of optimal and minimal conditions for population increase of some stored products insects. Journal of Stored Products Research 1:177–184.

IGLESIAS, H. A. AND J. CHERIFE. 1982. *Handbook of Food Isotherms: Water Sorption Parameters for Food and Food Components*. Academic Press, New York.

JACQUIN-DUBREUIL, A., C. BREDA, M. LESCOT-LAYER, AND L. ALLORGE-BOTTEAU. 1989. Comparison of the effects of a microwave drying method to currently used methods on the retention of morphological and chemical leaf characters in *Nicotiana tabacum* L. cv *Samsun*. Taxon 38:591–596.

KUCKAHN, T. S. 1771. Four letters from Mr. T. S. Kuckhan [sic], to the president and members of the Royal Society, on the preservation of dead birds. Philosophical Transactions 60:302–320.

LULL, W. P. AND B. P. MOORE. 1999. Herbarium building design and environmental systems. Pp. 105-118 in D. A. Metsger and S. C. Byers, editors. *Managing the Modern Herbarium.* Society for the Preservation of Natural History Collections, Washington, DC, xxii+384 pp.

MAYER, A. M., AND A. POLJAKOFF-MAYBER. 1975. *The Germination of Seeds.* Second edition. Pergamon Press, New York.

MICHALSKI, S. 1993. Relative humidity: a discussion of correct/incorrect values. Pp. 624–629 in *Preprints of ICOM Conservation Committee 10th Triennial Meeting.* Washington, DC, 22–27 August 1993. ICOM Committee for Conservation, Paris.

MICHALSKI, S. 1994. Leakage prediction for buildings, cases, bags and bottles. Studies in Conservation 39:169–186.

MILLER, R. B. AND A. RAJER. 1994. Freezing the herbarium: a novel approach to pest control. Poster summary, Preventive Conservation. Practice, Theory, and Research. International Institute for Conservation, Ottawa Congress, 12–16 Sept. 1994.

NANASSY, A. J. 1978. Temperature dependence of NMR measurement on moisture in wood. Wood Science 11(2):86–90.

NELSON, S. O. 1967. Electromagnetic energy. Pp. 89–145 in W. W. Kilgore and R. L. Roth, editors. *Pest Control, Biological, Physical, and Selected Chemical Methods.* Academic Press, New York. 477 pp.

NELSON, S. O. AND L. E. STETSON. 1974. Possibilities for controlling insects with microwaves and lower frequency RF energy. IEEE Transactions on Microwave Theory and Techniques, December:1303–1305.

O'NEILL, H. 1938. Heat as an insecticide in the herbarium. Rhodora 70:1–4.

OWEN, E. B. 1956. The storage of seeds for maintenance of viability. Bulletin 43, Commonwealth Agricultural Bureaux, Farnham Royal, UK.

PHILBRICK, C. T. 1984. Comments on the use of microwave as a method of herbarium insect control: possible drawbacks. Taxon 33:73–76.

PICHEL, W. 1965. Physical-chemical processes during freeze-drying of proteins. ASHRAE Journal 7(3):68.

PRIESTLEY, D. A. 1986. *Seed Aging, Implications for Seed Storage and Persistence in the Soil.* Comstock Pub. Co., London.

PYLE, M. M. AND R. P. ADAMS. 1989. *In situ* preservation of DNA in plant specimens. Taxon 38:576–581.

Rocha, F. F. 1959. Interaction of moisture content and temperature on onion seed viability. American Society for Horticultural Science 73:385–389.

Sinnot, Q. P. 1983. A solar thermoconvective plant drier. Taxon 32:611–613.

Stansfield, G. 1989. Physical methods of pest control. Journal of Biological Curation 1(1):1–4.

Strang, T. J. K. 1992. A review of published temperatures for the control of pest insects in museums. Collection Forum 8(2):41–67.

Strang, T. J. K. 1995. The effect of thermal methods of pest control on museum collections. Pp. 334–353 in C. Aranyanak and C. Singhasiri, editors. *Biodeterioration of Cultural Property 3*. Proceedings of the Third International Conference, Bangkok, Thailand, 4–7 July 1995. Thammasat University Press, Bangkok, x+718 pp.

Strang, T. J. K. and M. Shchepanek. 1995. Fast elevated temperature control of insects in vascular plant specimens. Poster, Society for Preservation of Natural History Collections annual meeting, 2–4 June, Toronto.

Toledo, R., M. P. Steinberg, and A. L. Nelson. 1968. Quantitative determination of bound water by NMR. Journal of Food Science 33:315–317.

Watling, R. 1989. Closing Remarks. Pp. 271–273 in N. Wilding, N. M. Collins, P. M. Hammond, and J. F. Webber, editors. *Insect-Fungus Interactions, 14th Symposium of the Royal Entomological Society of London*. Academic Press, London.

Whitehead, M. D. and G. F. Gastler. 1946–7. Hygroscopic moisture of grain sorghum and wheat as influenced by temperature and humidity. Proceedings of the South Dakota Academy of Science 26:80–84.

Womersley, J. S. 1981. Plant collecting and herbarium development, a manual. FAO Plant Production and Protection Paper, 33. Food and Agriculture Organisation of the United Nations, Rome.

Yamasaki, R. S. and A. Blaga. 1976. Hourly and monthly variations in surface temperature of opaque PVC during exposure under clear skies. Matériaux et Constructions 9:231–242.

SECTION II

ELEMENTS OF

HERBARIUM DESIGN

CHAPTER 5

Herbarium Building Design and Environmental Systems

WILLIAM P. LULL AND BARBARA P. MOORE

Abstract.—A herbarium, like any collections-holding institution, should provide a controlled environment in order to prolong the useful life of its collections. This involves many design features including architectural layout, structure, lighting, and fire protection. In addition, a herbarium should have special environmental systems to provide effective humidity control, to remove contaminants from the air and gaseous contaminants from the collection itself, and to maintain low temperatures to suppress insect activity and slow deterioration processes. If the collection includes wet specimens, a separate storage area should be planned with appropriate exhaust, fire detection, fire suppression, and perhaps with an explosion-proof design. As research into the long-term preservation of herbarium specimens and experience with herbarium environments increases, refinements to these guidelines may emerge.

INTRODUCTION

It has long been recognized that careful control of environmental factors such as humidity, temperature, light, and contaminants will significantly slow deterioration processes in many kinds of museum collections (Appelbaum, 1991; Thomson, 1986). Recent research work (Reilly, 1994; Sebera, 1994) with paper and other organic materials confirms this premise and demonstrates that the useful life of these materials can be significantly extended through storage at cool temperatures, low relative humidities, and with gaseous and particulate contaminants filtered from the air. To date there has been little specific research into the environmental needs of herbarium specimens. However, given the cellulosic nature of the majority of herbarium specimens, it seems reasonable to derive guidelines for herbaria from standards set for the archival storage of paper-based materials. As knowledge about the additional special environmental requirements of herbaria expands, further guidelines may emerge and the suggestions herein may need to be refined.

A herbarium construction or renovation project should aim to address most of the same environmental concerns as other institutions that hold collections—museums, libraries, and archives. In executing such a project it is essential to identify special environmental goals tailored to the needs of the collections, and to develop the project around these goals. The architecture, structure, lighting, environmental systems, and fire protection systems should then be developed from these goals, not adapted to them. Similarly, the design and construction team must be responsive to these goals, and the institution must take special care to check the designs against the goals to avoid mistakes in planning, construction, and operation. Only when these various aspects of a project are effectively addressed will a generally successful project result. Most unsuccessful projects can be traced to ill-advised compromises that lose sight of fundamental goals.

The following discussions highlight the more important aspects of the planning, architecture, and building systems that are needed to achieve a good herbarium preservation environment. Naturally, based on the ambient climate, projects can vary in both their architectural requirements and their equipment complements. For example, in consistently warm climates there may be no need for heating or humidification but rather an effective dehumidification system may be necessary. Lull and Banks (1990, 1995) provide a more detailed discussion of these problems as they relate to libraries and archives.

ARCHITECTURE

Some of the factors most critical to the success of a herbarium construction or renovation project involve the architectural layout and the building envelope design. These elements should complement and assist the mechanical equipment that is used to maintain a good conservation environment.

Space Segregation

Special attention must be paid to the environment in collection storage and usage areas. Therefore, segregation of the collection areas from non-collection areas that do not require this level of control is critical to an efficient and effective project. Division of space according to use allows concurrent segregation of the collections and reduction in the scope of special environmental controls. This, in turn, leads to associated reductions in the capital and operational expenses of the project.

Areas which might present a contamination risk to the collections should be physically separated from them with full-height walls and normally-closed doors. These areas include: offices, general supplies storage rooms, duplicating rooms, loading docks, and eating areas. Further, these areas must be on separate air systems to prevent cross-contamination. This arrangement will reduce project costs by allowing more conventional heating, ventilation, and air conditioning (HVAC) systems to be used in the non-collection areas.

Finishes

Room finishes in collection areas should be chosen to enhance lighting, to reduce short-term off-gassing of chemicals which can react with specimens, and to prevent long-term particulate contamination problems.

To optimize lighting, finishes of the walls, ceilings, shelving and other fixtures in collection storage areas should be white, and floors should be medium colored, not black. Alkyd and oil paints should be avoided due to their off-gassing. Rather, field-applied paints should be epoxy or acrylic. Manufactured fixtures such as herbarium cases and shelving should be supplied with a dry-process powder-coat finish rather than baked enamel, and case gaskets should be composed of a chemically stable material such as cellular silicone sponge, ethylene propylene diene monomer (EPDM), poly (dimethyl siloxane) (PDMS), or ethylene vinyl acetate (EVA).

Particulate contamination is controlled by ensuring that any exposed concrete or other cementitious material is properly sealed to consolidate the surface and prevent the sloughing of fine particulates.

Glazing

As a general rule, glazing and windows present many problems to collection areas. Glazing should be entirely eliminated from herbarium storage areas to prevent unnecessary thermal loads and light exposure.

Humidity-Tolerant Building Envelope

The building envelope may require special vapor control features to tolerate the humidified interior environment. In regions where the winter temperatures regularly drop well below interior dew point conditions, there is a risk of condensation on or within the building envelope when interior spaces are humidified, and appropriate precautions must be taken. The building envelope should be evaluated by the architect or engineer using the vapor diffusion model or other appropriate technique (ASHRAE, 1993). This quantitative evaluation of the wall construction should dictate the construction and location of any vapor barrier or insulation.

Water Risk Control

Measures must be taken to ensure that there is no chance of leaks from overhead construction or piping. Secondary protection is warranted where there is any risk. For example, if a roof is overhead without an intervening floor then the herbarium should have catch pans or shields that will ensure that water from any roof leaks will not reach inside the specimen cases. Any overhead piping with dynamic water flows (usually HVAC piping, but not sprinkler piping) should have catch pans beneath or should be double-sleeved.

The herbarium should not be subject to flooding. A basement location in areas where the water table is high is out of the question unless extraordinary measures are taken to control subsurface water.

LIGHTING

Lighting throughout a herbarium should be consistently good so that specimens in any location can be inspected with ease. Light damage to specimens is a function of illumination and time. Closed herbarium cases provide primary protection from light so light levels need not be kept as low as when organic collections are housed in the open. Even so, lighting treatments should be chosen to avoid excessive illumination levels and to remove the ultraviolet (UV) components while still meeting functional requirements.

Light Levels

Effective lighting in storage areas requires a minimum vertical illumination level of 2 foot candles (22 lux) to read labels at the worst location—usually the lower-most shelf. The most satisfactory means of attaining this is to employ indirect lighting or wide-distribution fixtures aligned 90 degrees to storage aisles. This orientation is particularly important where movable shelving is used.

Collection study areas need illumination levels of 40 to 80 horizontal foot candles (440 to 880 lux) at the working plane. Lower levels may not provide adequate illumination for the necessary inspection of specimen detail. Higher levels can cause unnecessary damage to collections.

It can be problematic to rely on natural daylight to illuminate work areas. On dark days the light level is too low while on bright days it is too high, so relying on daylight as a light source is not recommended.

UV Control

The ultraviolet content of light is especially damaging to specimens. The best way to control UV exposure is to select light sources that are inherently low in UV emissions, such as tungsten and certain fluorescent lamps. Where light sources emit higher or variable UV components, then light sources must be filtered. Most UV filters are applied at the light source, for example a sleeve filter for the typical fluorescent tube lamp. UV can also be filtered from light by making the light entirely indirect, as by bouncing it off a white surface painted with a titanium dioxide pigment. However, remember that even when the UV content is completely removed from light, the visible light itself can do considerable damage to a collection. Exposure even to filtered light should therefore still be limited. Specimens that are not being worked with should always be returned to cases. Further discussion of lighting is provided in Lull (1982), Lull and Merk (1982).

HVAC SYSTEMS

The special environmental needs of herbarium collections can be met through the use of a well-designed heating, ventilation, and air conditioning (HVAC) system. The system should provide effective cooling, humidification, dehumidification, filtration for particulates and gaseous contaminants, and air circulation to suppress mold growth. It should be reliable, and reasonably economical to run and maintain.

Humidification equipment should be chosen for good part-load control and low maintenance, and should avoid designs based on reservoirs of standing water near room temperature.

Key features that should be specified for HVAC systems designed to serve collection spaces are:

- twenty-four hour operation—so that stable conditions are maintained;
- constant air flow—to assure even mixing and filtration of room air, and to suppress mold growth;
- ducted air distribution with forced-air heating;
- cooling;
- positive dehumidification—usually cooling with reheat capability;
- humidification—usually using clean steam;
- particulate control—to prevent recirculated contamination of the collection spaces and HVAC distribution system;

- gaseous contamination control, as warranted.

Woods (1983) provides specific environmental and system criteria for paper-based collections which should, in large part, apply to herbaria.

Air Delivery

Collection environments rely on air flow to filter damaging pollutants and contaminants, to control humidity for the collection, to lower any potential for mold growth, and finally to heat and cool the collections space. Most institutions have found these needs met best by constant-volume all-air systems with central air-handling stations and dedicated ducts for each collection space. Filtration, dehumidification, humidification, and maintenance should be in a central location away from collections to avoid potential leaks from water pipes and risks associated with maintenance in the collection areas.

Variable volume (VAV) systems, which are appropriately used for office environments, are more economical to install and to operate than constant-volume systems, but cannot easily be modified to achieve a good collection conservation environment. Attempts to do so have led to problems in system balancing, an inability to hold winter or summer humidity, and problems with overhead reheat water and steam piping. In virtually every case, the cost and space required for the properly designed VAV system (full filtration, local humidification, local dehumidification, minimum air volume settings, well-planned piping, well-planned maintenance access, and extremely well-documented operating instructions) gives no advantages over the more conservative, less problematic, and easier to maintain constant volume systems. Thus, while the VAV system can be made to save energy compared to the constant-volume system, it does so at the expense of the collection. There are many options and features available for a constant volume system that can achieve all but the fan energy savings of a VAV system—fan energy that is required to maintain air circulation.

Temperature

Recent research (Reilly, 1994; Sebera, 1994) shows that the life of paper and other organic materials can be extended by storage in cooler, drier conditions. Therefore, assuming that plant material deteriorates in a manner similar to paper, lower temperatures should extend the life expectancy of herbarium specimens. In addition, many insects that attack herbarium specimens are dormant at lower temperatures, generally below 60°F or 65°F (17°C or 19.5°C; Strang, 1992). Thus it follows that the most suitable mechanical systems for herbaria are those that are able to maintain temperature levels low enough to control those insects which are thought to be a threat.

Humidity Control

Sealed or semi-sealed cases provide excellent buffering of minor humidity fluctuations shorter than 24 hours in duration. Thus, where herbarium specimens are enclosed in cases, humidity cycling and minor short-term humidity fluctuations need not be considered a problem. Packaged HVAC system components, such as direct-expansion (DX) cooling equipment, that typically cause minor rapid humidity fluctuations, can therefore be used in a herbarium without the concern that might pertain if specimens were stored on open shelves. Humidifiers with non-pressurized electric canister boilers or evaporative pans may also prove acceptable since the humidity cycling typical of these systems is minor and rapid.

The best mechanical systems for herbaria should maintain an average humidity below 50% RH. It is critical to avoid any chance that herbarium collections will be subjected to prolonged exposure to high humidities. This requires positive dehumidification, over and above the coincidental dehumidification that can occur with most cooling systems. Positive dehumidification usually involves a cooling coil and a reheat coil with the proper controls.

Research (Reilly, 1994; Sebera, 1994) indicates that humidity as low as 30% RH may dramatically extend the life of paper-based collections. Although the research did not specifically include herbarium specimens, given the adhesives used with herbarium specimens, a humidity goal of around 40% RH may be a good choice. In some circumstances this lower humidity may require the use of a desiccant system. In a desiccant system, part of the recirculated air is passed through the desiccant for additional drying, as a supplement to (not a substitute for) a cooling coil/reheat coil arrangement. Desiccants have long been used for effective dehumidification in film storage vaults. They are now being used to dramatically extend the life of paper-based collections in libraries and archives, and can be used to similar ends for herbaria. Backup or redundant systems may be indicated where system failure scenarios predict loss of dehumidification for a time period that would allow specimens to respond to a higher humidity.

Gaseous Filtration

In cases where collections have in the past been aggressively treated with insecticides, the threat of gaseous contamination from the collection itself is greater than that from outdoor pollution or ambient contamination. In these cases gaseous contamination control is designed not only to remove outdoor pollution, but, more importantly, to remove the various volatile insecticides that may both do long term damage to treated and adjacent specimens, and pose a health risk to herbarium users. The chemical media used in such filtration systems should be carefully selected to be effective on the target contaminants. The gas filtration system should be designed to ensure easy maintenance and easy checking for expended media.

FIRE PROTECTION

Unlike HVAC and lighting systems, fire protection systems do not have an active daily impact on a collections conservation environment, but their features and function can be critical to the safety of the collection. Effective fire protection requires early detection and effective suppression.

Detection

The fire detection system should allow sensitive and early detection of any potential fire condition. This means that detection should be as broad as possible, relying on both obscuration (photocell) and combustion by-products (ionization) detectors. The fire detection system for storage spaces housing wet specimens in alcohol must be capable of detecting smokeless alcohol fires. This usually requires use of ionization detectors. To confine the response to specific detector locations, the detectors should be individually addressable rather than grouped into zones. To ensure early detection the detectors should send an analog signal to a monitoring panel, instead of a smoke/no smoke signal. The analog signal allows the sensitivity of individual detectors to be adjusted for their expected nuisance contamination conditions. This means a detector near the loading dock might be set to a high alarm threshold, and a detector in a collection storage room set to a much lower level.

Suppression

When considering fire protection design, it is important to remember that the first priority for design is life safety, and that will be the primary, and possibly exclusive, goal of the local fire codes and code officials. In many cases the fire protection codes are intended to slow the spread of fire, and not necessarily to protect the contents of the building. The goal is often to buy time, rather than to protect property. This is why features that provide additional protection to the collection, beyond the fire code, may be necessary in a herbarium.

By far the most proven and reliable method of fire suppression is water sprinklers. In some instances, fire codes may allow the option of manual suppression, such as with fire extinguishers and water hoses, rather than automatic sprinkler systems. The delay in response of manual suppression may be an acceptable risk in some circumstances, but considerable damage to a collection may be done in the time it takes to mount a manual suppression effort. The institution should also seriously consider the damage that a high-pressure water hose, which can literally cut through drywall partitions, may do to the collection. In many instances the threat posed by a well-designed automatic sprinkler system is considerably less serious than the threat posed by a manual system in the course of fire suppression.

The most common and least expensive type of sprinkler system is a wet-pipe system. In this type of system the protected space has piping overhead charged at all times with water under pressure (hence, wet-pipe). A fire causes a sprinkler head to open and discharge water. The only real disadvantage of this type of system is the threat of accidental water release when there is not a fire. This can occur if a head is damaged or, perhaps more commonly, following accidental freezing of pipes.

The threat of freezing can be addressed through the use of a dry-pipe system, where the water in the piping is held back by compressed air and a pressure-dependent valve. Once a sprinkler head opens for any reason the compressed air escapes, and the water is released.

A third option is a pre-acting sprinkler system. In these systems, water is not present in the sprinkler piping until several conditions call for the release of water into the system. The conditions required include the presence of smoke, fire, heat and/or the opening of a sprinkler head. Water is not discharged into the space until a sprinkler head has been opened by space temperature, and after the smoke and/or heat conditions have triggered the pre-action of releasing water into the piping. This type of system affords protection from leaking sprinkler piping or damaged sprinkler heads as well as from freezing. This system is more expensive than a wet system owing to its complicated sensors and controls. It requires more maintenance while providing a slightly lower level of fire safety. It also requires multiple smoke detectors in each protected space, which may be an added expense.

In non-storage, collection support areas such as offices and lab/ work areas, the reliable wet system is usually the appropriate choice. For storage areas in which the collection has a primary enclosure, such as well-sealed herbarium cases, a wet system might also be indicated. However, in or above collection areas where specimens are not in cases and would be damaged by an accidental discharge, a wet system could constitute an added risk. Each institution must balance, according to its own circumstances, the risk of exposure of the collection to a modest amount of water against the installation cost and annual maintenance premiums of a pre-acting system.

Gaseous Fire Suppression Agents

Gaseous fire suppression agents may seem an attractive alternative to sprinkler systems since such systems promise to extinguish a fire without introducing water. However, gaseous suppression agents have proven very expensive, and are not nearly as reliable as water sprinkler systems. Other problems include difficulty in maintaining gas containment, threat from breaching fires, cost of false discharges, need for involved annual testing, diminished availability of suppression agents (such as halon), high cost for substitute agents, and the ultimate need to back up any gas system with water sprinklers. In addition, if National Fire Protection Association (NFPA) regulations are followed by the fire department, after a fire has been suppressed by a gas system, the collections areas near the site of the fire will be hosed down with water. Thus the primary perceived benefit of a gas fire suppression system is lost. For all these reasons fire suppression should not be based on gas or other unproven systems; a water sprinkler system should always be preferred.

WET SPECIMENS

Wet specimens in formaldehyde or alcohol require different conditions than other herbarium collections and so must be isolated from other collections storage and usage areas. The temperatures in wet storage areas should be low to inhibit the evaporation of the preservation solutions. The humidity should also be maintained at moderate levels, as high relative humidity presents an indirect risk to the specimens over time due to differential evaporation of the preserving solution components. Since the specimen solutions release small amounts of formaldehyde or alcohol over time, the space must have an appropriate continuous exhaust. In some cases building codes or good engineering practice may dictate the use of explosion-proof design for lighting, equipment and wiring devices.

If the wet specimens are in closed cases, these may need to be vented or connected to the exhaust system. If the specimens are stored on open shelves, room lights must have a low ultraviolet content, light levels must be low to reduce the cumulative light exposure of the collection, and lights should be on only when necessary.

COSTS, CONSTRUCTION, AND OPERATION

A herbarium should be designed to ensure the good long-term preservation of the collection. Attention to special herbarium needs, as described, can lead to savings in some areas and greater expenses in others. However, by defining the needs early in a building or renovation project, the institution stands the best chance of those needs and concerns being met in the herbarium construction project. The actual construction should be closely monitored and inspected and new systems should be broken in and commissioned before occupancy. The primary functional goals of the collection environment should be reflected in the operation of the facility.

LITERATURE CITED

APPELBAUM, B. 1991. *Guide to Environmental Protection of Collections.* Sound View Press, Madison, Connecticut.

[ASHRAE] AMERICAN SOCIETY OF HEATING, REFRIGERATING AND AIR-CONDITIONING ENGINEERS, INC. 1993. *Handbook: Fundamentals.* I-P Edition. Atlanta, Georgia.

LULL, W. 1982. Preservation Aspects of Display Lighting. Electrical Consultant November/December 1982:8-21, 39.

LULL, W. AND L. MERK. 1982. Lighting for storage of museum collections: developing a system for safekeeping of light-sensitive materials. Technology and Conservation 282:20-25.

LULL, W. AND P. N. BANKS. 1990. *Conservation Environment Guidelines for Libraries and Archives.* New York State Library.

LULL, W. AND P. N. BANKS. 1995. *Conservation Environment Guidelines for Libraries and Archives.* Canadian Council of Archives.

REILLY, J. 1994. *New Tools for Preservation: Assessing Long-Term Environmental Effects on Library and Archives Collections.* Image Permanence Institute, Rochester Institute of Technology, Commission on Preservation and Access, June 1994.

SEBERA , D. 1994. *Isoperms: An Environmental Management Tool.* Commission on Preservation and Access, Washington, DC.

STRANG, T. J. K. 1992. A review of published temperatures for the control of pest insects in museums. Collection Forum 8(2):41-67.

Thomson, G. 1986. *The Museum Environment.* Second edition. Butterworths, London.

Woods, J., editor. 1983. *Air Quality Criteria for Storage of Paper-Based Archival Records.* National Bureau of Standards, NBSIR 83-2795.

CHAPTER 6

Elements of Herbarium
Layout and Design

BARBARA ERTTER

Abstract.—The layout and design of an herbarium can significantly affect efficiency of operations. Many of the factors critical to an efficient design may be easily overlooked in the early planning stages of an herbarium renovation or construction project. A compilation of these factors is presented, drawn primarily from personal experience in herbarium renovation at the University of California at Berkeley. The focus is on those aspects of herbarium design for which there is a dearth of existing references. Emphasis is placed on the quality of the work environment and the importance of planning the layout of work spaces to ensure efficient flow of specimens as they are processed. No universally perfect design is suggested. Rather, a checklist of factors is presented for consideration and to balance against existing constraints and preferred techniques in each herbarium.

INTRODUCTION

A well-designed herbarium will increase the efficiency and efficacy of curatorial operations, research access, pest control, and collections maintenance in general. There is no universally perfect design; rather, there is a unique solution for each herbarium, based on an equation that considers and balances multiple factors.

Whether constructing a new building, or renovating an old one, many of these design factors are dictated by inflexible parameters such as the size and shape of the available space, local building codes, budgetary constraints, and available staffing. Making the best use of those factors that remain flexible within these constraints requires careful analysis of and planning for actual herbarium functions. This involves many critical details that can be overlooked even by experienced herbarium curatorial staff.

Any major herbarium construction or renovation will involve professional architects. The focus of this paper is therefore on those factors that are outside the expertise of most architects, particularly those that are not dictated by local codes and for which there is a dearth of published references. Much emphasis is placed on the desirable proximity of work spaces to optimize the efficient flow of specimens, supplies, and people, both staff and visitors, in the course of routine herbarium operations. Desirable proximities are diagramed in Figure 1.

The following compilation of the various factors that should be considered when planning an herbarium is based primarily on my personal involvement with designing, moving specimens into, and subsequently using four herbarium facilities over a twelve-year span: the herbaria at the University of Texas at Austin (TEX and LL) in 1984, and the University (UC) and Jepson (JEPS) Herbaria of the University of California at Berkeley between 1986 and 1995 (Ertter, Chapter 9, this volume). The examples are largely drawn from the latest move, that of UC and JEPS, into renovated quarters in the Valley Life Sciences Building (VLSB; Fig. 2).

Figure 2 shows the initial floor plan of the renovated UC-JEPS, in which some areas have already been modified for other functions (e.g., the types vault was converted to general storage when leaks became apparent). This floor plan provides not only an example of one unique solution to the balance of desirable proximities (e.g., specimen handling areas), but also several shortcomings (e.g., libraries separating the reception area from research offices and break/conference room; archives placed near laboratory containing potential hazards). Some flaws were not evident until occupancy, while others inevitably resulted from the impossibility of obtaining all desired goals simultaneously, especially

FIGURE 1. Schematic of desirable proximities of herbarium areas, as indicated by connecting lines. Connections through nodes are equivalent to uninterrupted straight lines.

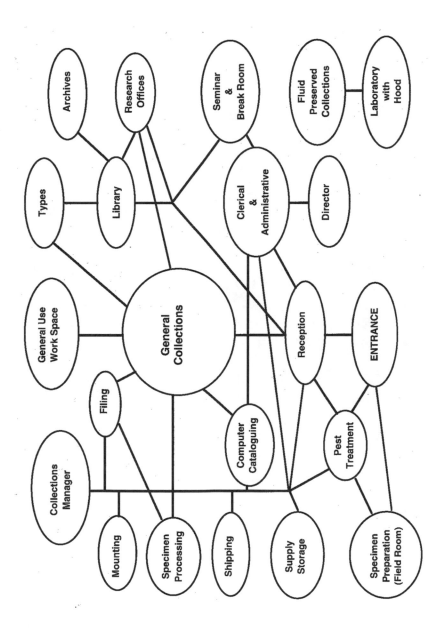

Figure 2. Floor plan of UC-JEPS Herbaria in renovated Valley Life Sciences Building.

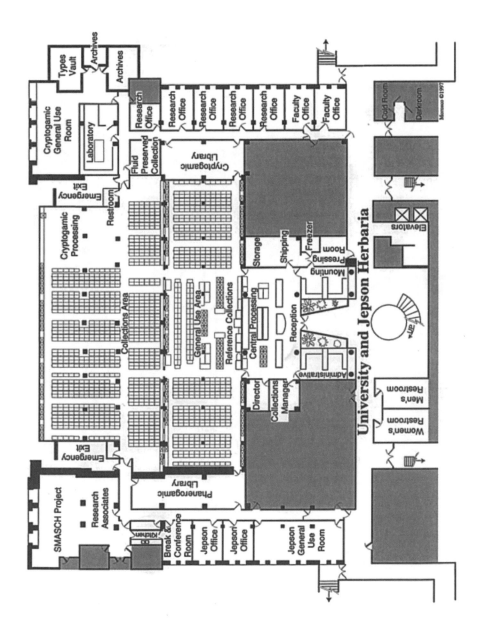

within the constraints of an existing architectural footprint. In spite of these recognized deficiencies, the UC-JEPS lay-out has proved to be highly functional to both staff and visitors.

The recommendations presented here are based on both positive achievements and recognized deficiencies in the renovated spaces.

A. GENERAL CONSIDERATIONS

1. Work Environment

The work environment can play a major role in herbarium function. Poor design features negatively affect the comfort and health of herbarium personnel, contributing to increased stress levels, lowered daily productivity, and increased sick leave.

Factors to consider:

- *natural lighting:* most likely to be important for support staff who work in a single area for long periods of time. In UC-JEPS, a light well in the atrium provides natural lighting to interior work areas.
- *ergonomic design:* particularly at stations where work continues for prolonged periods, and if constant repetitive tasks are involved (e.g., mounting, microscope, and computer stations).
- *potential irritants* and *perceived dangers* (e.g., persistent noise, chemical odors, electro-magnetic radiation): some of these may not always be evident until after moving in. Take the concerns of staff seriously; they may serve as a thermometer to environmental safety.
- *wheelchair accessibility:* at least for some work stations and primary access routes, particularly those leading to emergency exits. The entire collection at UC-JEPS did not need to be wheelchair accessible, as long as assistance was provided when needed.
- *staffing:* work spaces for tasks performed by the same person or by back-up staff (e.g., specimen handling and clerical functions) should be in close proximity. The relevant tasks will be unique to each institution.

2. Pest Control

Control of insect pests in herbaria and other natural history collections is a rapidly evolving field with an expanding literature (Hall, 1988; Linnie, 1990; Strang, Chapters 3 and 4, this volume) and was also the topic of a 1992 SPNHC workshop "Getting the Bugs out of the System: Museum Pests from Bacteria to Beetles." For a variety of reasons, including human health (Linnie, 1990), the heavy reliance on chemical controls is giving way to integrated pest management (Strang, Chapter 3, this volume). Only those strategies directly involving facilities design or use are discussed here. The primary herbarium pests are the cigarette beetle (*Lasioderma serricorne*) and the drugstore beetle (*Stegobium paniceum*), members of the Anobiidae.

Strategies for pest control:

- *provide a quarantine area for untreated specimens.*
- *maintain the remainder of the herbarium as a pest-free environment.* At UC-JEPS, all specimen processing, collections storage, and research offices comprise a single bounded space within which treated specimens can be freely transported. No dried specimens are allowed in without first undergoing treatment, and any specimens removed from this bounded space, even temporarily (e.g., for class use), are treated prior to re-entry.
- *isolate the collections area* from other parts of the herbarium by well-sealed walls to provide an additional physical barrier to pests and to allow the collection to be maintained at a lower temperature for pest control purposes and/or archival climate control.
- *keep the number of doors to a minimum.*
- *keep doors shut.*
- *block gaps under doors with door sweeps* to discourage crawling insects.
- *install screens of 18-20µ mesh on unsealed windows.*
- *seal holes in walls, ceilings, and floors:* these are potential entry and breeding sites for insect pests.
- *install filters in ventilation systems.*
- *be aware that water sources* (e.g., sinks, building water pipes in-

cluding those hidden behind suspended ceilings) *can create a local area of higher humidity that provides habitat and a breeding ground for insects.*

3. Potential Hazards

The collections area (especially the type collection), library and other archival resources should, where possible, be separated from potential hazards. While this is common sense, the challenge is to realize what counts as a potential hazard. Laboratories and kitchens constitute both fire and flood hazards. *The wisdom of locating laboratory or food preparation facilities next to irreplaceable collections, no matter how convenient, should therefore be carefully evaluated.*

Fire hazards:

- *cardboard and other flammable items stored next to a drier* or any other potential source of ignition.
- *unvented plant driers:* all driers should be adequately vented, especially if they might be used for material that has been initially preserved in alcohol or formalin (e.g., vascular plants collected in the tropics).
- *kitchen appliances:* toaster, kettle, hotplate, etc.

Flood hazards:

- *laboratory safety shower*
- *automatic sprinklers*
- *fire fighting activities*
- *water sources in general* (sinks, water pipes including those hidden behind suspended ceilings)
- *damp corners and floor drains*, especially in subterranean locations, can cause serious seepage or flooding problems.

Seismic hazards:

- *unanchored bookcases and other tall free-standing objects*, especially those located along exit routes. When anchoring storage equipment, avoid drilling holes into herbarium cases that will compromise their insect-proofness. At UC-JEPS, standard her-

barium cases danced slightly but did not tip over in the 7.1 Loma Prieta earthquake.
*   *open shelves above regular work sites*

4. Security

While herbarium collections generally do not have the same black-market value as do many paleontological, anthropological, and zoological collections, security factors should nevertheless be given careful consideration. Computers, library materials, and personal belongings can be attractive to thieves. In addition, the personal safety of people working in the herbarium should always be a high priority.

Security strategies:

*   *restricted access into the herbarium.* Because UC-JEPS was designed as a single bounded unit, entryways could be restricted to a single staffed reception area for public access with a minimum of other doors (which are kept locked at all times). Two doors opening into secluded alcoves, required as emergency exits, were fitted with alarms. Ground level windows, some of which abut secluded window wells, have been an additional security concern.
*   *hierarchical keying systems achieve desired security levels.* At UC-JEPS, the keying system was planned to allow four levels of access: *level one:* main collections and break room only; *level two:* level one access plus non-occupied support areas, including the library and laboratory; *level three:* all areas of level one and two access plus rooms occupied by several people, each such room keyed separately; *level four:* herbarium master. There are also separate keys for individual offices, primarily for allowing access by collaborators. With this system, most staff need to carry only one key, with allocation determined by need. The same strategy can be achieved with magnetic swipe cards.

5. Visitors

Even for an herbarium with no public access, consideration should be given to how visitors (and delivery personnel) are to announce their arrival and gain access. The flow of visitors from reception is also important. This is particularly critical if faculty, who need to be accessible to students, have offices located inside the herbarium. Routes to areas frequently accessed by visitors (e.g., staff offices, seminar rooms) should either be obvious and/or clearly marked, or else the visitor will need to be escorted. Casual visitors should not be routed through special security areas (e.g., libraries) without an escort. One flaw in the UC-JEPS design is that the most direct access route to research staff offices is through the library (Fig. 2).

A system should be devised for notifying visitors in the event of phone calls, providing access to photocopiers, and otherwise accommodating them during their stay. This includes access to restroom facilities when doors are locked.

B. DESIGN CONSIDERATIONS FOR COLLECTIONS AREAS

1. General Collections

The general collections are the heart of the herbarium. Specimens are most often housed in standard size herbarium cases, which can be placed in free-standing rows or on moveable aisle compactor systems. In North America, herbarium cases are commonly 29-30 inches (74-76 cm) wide, 19 inches (48 cm) deep, and either 7 or 7.5 feet (2.13 or 2.29 m) tall, with a single door 26-28 inches (66-71 cm) wide. Double-door versions 55 inches (140 cm) wide are also available. These dimensions often determine the most efficient arrangement of rows and aisles.

Location factors:

- *size of the collections*
- *strength of the floor* (including compact storage system if used). Floor loading capacity is discussed by Rabeler (this volume). Many herbaria, including UC-JEPS, are located on the ground floor where floor loading capacity is not a problem, in spite of the potentially increased flooding hazard. The weight of the cases must also be considered when calculating the floor strength. Based on the case sizes mentioned above, an empty single case weighs 350 lbs. (159 kg) and an empty double case weighs 550 lbs. (250 kg). The weight of the loaded case may be calculated by multiplying the weight of specimens in one pigeonhole — approximately 6 lbs. (2.72 kg) —

by the number of pigeonholes in a cabinet. The working weights used for the renovation of the University of Michigan Herbarium (MICH) were 500 lbs. (227 kg) per loaded single case and 850 lbs. (386 kg) per loaded double case (Richard Rabeler, MICH, pers. comm.).

- arrangement of herbarium cases that best allows *uninterrupted left-to-right filing sequence, with particular taxonomic groups located near corresponding research offices or facilities* (e.g., mycology near laboratory space).
- constraints imposed by *existing architectural footprint and features* (e.g., overhead pipes lower than height of standard herbarium cabinet; control panels that can't be blocked).

Equip with:

- *good lighting over aisles and work areas* (see Lull and Moore, this volume; Rabeler, this volume)
- *surfaces on which to place specimens for quick examination and filing:* place throughout the collection.
- *work stations (Fig. 3) for more detailed specimen examination:* can be in an adjacent room if the collection is maintained at a temperature below comfort level. Each station should be equipped with:

FIGURE 3. Work stations in UC-JEPS collections area.

- *dissecting microscope* with clearance for the boom arm
- *illuminator*
- *electrical outlet*
- *space for herbarium tools:* orientation guides, annotation supplies, and basic tools (e.g., dissecting needles)
- *empty half-height case* to store specimens
- *overhead shelf* for reference materials and personal belongings (optional)
- *means of communicating* with people in the collections area should be provided, especially in a large collection.

Provide access to:

- *networked computer and printer*
- *appropriate reference materials* (e.g., local floras)
- *rolling carts*
- *copy stand* for photographing specimens
- *photocopy machine*
- *filing supplies* (perhaps at a mobile station)
- *paper cutters*
- *typewriter*
- *bulk storage of folders:* may need to be in a separate general storage area.

2. Special Collections

In addition to the main collection, space must be allocated for various special collections, each of which may have its own requirements.

- *types:* require extra protection, possibly controlled access.
- *synoptic collections:* convenient for responding to public service requests.
- *controlled substances,* e.g., *Cannabis,* which have special legislative requirements.
- *material on loan from other institutions:* for oversized specimens, e.g., those from some European herbaria, use a generic museum case, twice as deep as a standard herbarium case and with removable drawers — available as a catalog item from most case manufacturers.
- *specimens in various stages of processing*

3. Fluid-preserved Collections

Most features of the storage area intended for fluid-preserved collections will be dictated by local and institutional codes for environmental health and safety, as well as planning for emergency situations such as earthquakes. Alcohol is flammable, while formaldehyde is a potential carcinogen. Primary considerations are adequate venting and potential ignition sources (e.g., lighting systems). The City of Berkeley required an estimate of the total amount of formaldehyde. Our original figure for UC-JEPS, based on the number and volume of containers, plummeted when we considered the dilution factor and the actual volume in containers, including displacement by specimens.

Place close to:

- *fume hood* for formalin-preserved specimens

Equip with:

- *shelves with lips* or some other means of keeping containers from falling off, in the event of tremors in earthquake-prone areas. At UC-JEPS, standard metal utility shelving has been modified for this purpose (Fig. 4).

FIGURE 4. Shelving for fluid-preserved collections in UC-JEPS.

- *additional containment contingencies* (e.g., a bag of kitty-litter for spills is a simple starting point). In VLSB, all fluid-preserved storage rooms drain into a special underground storage tank, to avoid contaminating the local water treatment system in the event of a major spill. As a result, it will be difficult to retroactively convert additional space for the storage of fluid-preserved collections.

C. DESIGN CONSIDERATIONS FOR SPECIMEN HANDLING AREAS

The efficient flow of specimens is a primary concern in designing herbaria. Areas where specimens and related supplies are sequentially processed and stored should be in close proximity. Specimen processing and storage areas should be designed so that the layout follows the sequence of tasks to be performed, from the pressing and drying of freshly collected material, through treatment for pests, to counting, cataloging, identifying, mounting, accessioning, labeling, and filing. It may not be possible to arrange all processing areas adjacent to each other, so it is important that materials and specimens can be transported conveniently between areas on rolling carts.

In general:

- *base the dimensions of work surfaces, shelves, and storage areas on multiples of the dimensions of standard herbarium sheets, supplies, and cases* (in North America, 12 ins. x 18 ins. [30.5 cm x 46 cm] is considered standard). This will avoid sheets hanging off surfaces.
- *provide sufficient case space in which to house specimens at every stage of specimen processing to prevent pest infestation.* One innovation at UC-JEPS is a designated case used to house specimens being routed to someone. A card inserted in the person's mailbox indicates that a labeled bundle of specimens for their attention can be retrieved from the routing case. This system avoids leaving specimens unattended on the desk of an absent researcher.
- *avoid the need for routine transport through the central collections area* so as to reduce the possibility of spreading pests, especially from untreated material.

- *avoid stairs* as they act as a major barrier to transporting specimens by preventing the use of carts.

1. Pest Treatment Area

Separate from:

- *main collection*, as this area serves as quarantine quarters.

Place close to:

- *reception area*
- *specimen processing*

Equip with:

- *freezer or alternate pest treatment equipment* (e.g., heat or fumigation chamber): the equipment should be large enough to contain the contents of an entire herbarium case, thereby allowing the case to be emptied and cleaned in the event of an infestation. In VLSB, a walk-in blast freezer is located between the specimen preparation/quarantine area — which has a separate entrance — and the main specimen processing area.
- *holding case:* to accumulate miscellaneous bundles of specimens between treatment cycles, since opening a freezer in the middle of a cycle can undermine the effectiveness of the treatment.
- *microwave oven:* for treating specimens brought by visitors; big enough to accommodate herbarium sheets. [*Editors note: this practice is not universally accepted (see Strang, Chapter 4, this volume).*]

2. Specimen Preparation Area

The specimen preparation area (pressing room, field room) is where freshly collected materials are pressed, dried, and otherwise initially processed.

Separate from:

- *areas where treated specimens are processed and stored:* although the majority of insects that might be on fresh material (e.g., caterpillars) are of no threat to dried plant specimens, there is nevertheless the risk of introducing an infestation. At UC-JEPS, a field or pressing room with a separate entrance serves as a preparation area for freshly collected material and also as a quarantine area for any incoming material prior to treatment.

Place close to:

- *pest treatment area*
- *shipping area:* many supplies (corrugates, newsprint) are used for both preparation and shipping of specimens.

Equip with:

- *plant press drier:* either commercially produced or constructed in-house. It should be adequately vented, especially if used to dry formalin-preserved specimens.
- *separate drier for fleshy fungi:* if fungi are routinely dried at a different temperature.
- *counter space*
- *sink with a dirt trap:* if the sink is also used for sorting buckets of marine algae preserved in dilute formalin, it should be relatively shallow and wide, and be fitted with active ventilation (e.g., laminar flow hood).

Provide storage for:

- *presses and related preparation equipment*, with user access and restrictions taken into consideration.

3. Specimen Processing Area

The specimen processing area (Fig. 5) is where incoming and outgoing loans, exchanges, gifts, and newly mounted material are sorted, counted, and otherwise processed. In large herbaria, satellite sorting areas might also be useful, as well as separate offices for collections management staff. There is furthermore a major clerical component, involving the various supplies and resources discussed in the section "Design Considerations for Clerical, Administrative and Reception Areas".

Separate from:

- *areas in which untreated specimens are stored*

Figure 5. UC-JEPS specimen processing area.

Place close to:

- *collections storage*
- *pest treatment, mounting, cataloguing, shipping, clerical area*
- *collections management staff offices*

Equip with:

- *ample work surfaces, both standing- and sitting-height:* deep enough, at least 36 ins. (90 cm), for a double row of specimens. L- and U-shaped counters can provide efficient work surfaces, perhaps created from pull-out shelves.
- *banks of sorting shelves:* (optional) those at UC-JEPS are adjustable.
- *herbarium cases:* for storing specimens in various stages of processing so that no specimens are left out overnight. Back-to-back rows of half-height cases are a logical answer to multiple needs; a counter placed on top will minimize the accumulation of pest-harboring debris between cases.
- *storage space:* for appropriate processing supplies, records and reference materials, and for rolling carts when not in use.

4. Mounting Area

The mounting area is where plants are mounted, boxed, packeted, or otherwise converted into fully-prepared specimens. The optimum arrangement of the space will be heavily influenced by the mounting technique used.

Place close to:

- *specimen processing, pest treatment, cataloguing, shipping, and collections storage areas*

Equip with:

- *work surface* sufficient for several stacks of herbarium sheets, supplies and equipment at each station: L- or U-shaped configurations work well (Fig. 6).
- *above-counter shelf:* for supplies and equipment needed while mounting (optional): make sure there is sufficient clearance between counter and shelf for the stacks of specimens.
- *storage for working quantities of supplies* such as mounting paper and glue.

FIGURE 6. UC-JEPS mounting area.

- *sink:* needed for most mounting techniques, if only to clean equipment.
- *herbarium cases:* to house specimens before and after mounting, possibly with one case or section of shelves reserved for priority mounting, repair and/or other special treatment.

Provide access to:

- *photocopy machine:* for duplicate labels (on acid free paper)
- *paper cutter*
- *bulk supplies:* should be easily transportable by rolling cart or hand-truck (i.e., no stairs).

5. Shipping Area

The shipping area is where specimens are wrapped for shipment. The contents and arrangement of the work space are determined by wrapping technique and available equipment (e.g., rolls of wrapping paper). At UC-JEPS, both high and low counters are available for different tasks and to suit personal preferences (Fig. 7).

Place close to:

- *specimen processing*

FIGURE 7. UC-JEPS shipping area.

- *clerical areas:* many of these supplies are generated as incoming shipments are processed.
- *bulk storage*

Equip with:

- *ample storage space:* for boxes of various sizes, cardboard, padding, and other shipping supplies.
- *work surfaces*

6. Cataloguing and Data Entry Area

Place close to:

- *related specimen processing*
- *clerical areas*

Equip with:

- *computer(s), printer(s) and related equipment*
- *ample electrical and computer outlets:* for computers used for cataloguing.
- *dedicated surge-protected lines and emergency power*
- *adequate work surface for stacks of herbarium specimens*
- *herbarium cases* in which to store the specimens.
- *shelves for appropriate reference materials*

D. DESIGN CONSIDERATIONS FOR CLERICAL, ADMINISTRATIVE AND RECEPTION AREA(S)

1. Reception Area

In addition to controlling access, the reception area is where visitors are greeted, screened, and oriented. Mail and other deliveries can be received and picked up, and incoming dried specimens immediately diverted for pest treatment. Staff from this area may be needed to escort or announce visitors.

Place close to:

- *collections, offices, library*
- *back-up personnel:* in herbaria with public access, particularly

those with active public service and outreach programs, reception becomes a high staffing priority. At UC-JEPS, collections management personnel routinely serve as back-up to the receptionist at an entrance open to the general public (Fig. 8). These two functions are therefore in shared space.

- *pest treatment*

Figure 8. Combined reception and specimen processing areas in UC-JEPS.

Equip with:

- *routing information for visitors*
- *visitor register and any orientational or instructional handouts*
- *informational brochures and displays:* this is particularly important when financial support may depend on effective outreach.
- *a sturdy reference collection of local plants* in the reception area (optional).
- *dissecting microscope:* for public service identification of specimens, perhaps at a work station that could also be used by visitors needing supervision.
- *some basic reference books and/or a computer terminal* providing access to collections-related information (e.g., the herbarium

home page): even if public service queries are generally routed to other personnel. These may include guides to the local flora, plant names, poisonous plants, medicinal plants, and cultivated plants.

- *telephone equipment*
- *electrical outlets*

2. Clerical and Administrative Area(s)

Although administrative functions (e.g., personnel matters) might not occur in the same area as general clerical functions (e.g., preparation of filing of shipping forms), and may actually occur in departmental offices outside of the herbarium proper, there is enough overlap in responsibilities and requirements that they are discussed here as a unit. Thus it may be desirable to locate all staff offices in this area, if only to allow shared use of office equipment and supplies.

Equip with:

- *photocopy machine*
- *printers*
- *fax machines*
- *paper cutters*
- *clerical and computer supplies*
- *networked computers*
- *telephones*
- *typewriters*
- *pencil sharpeners*

Include necessary electrical, phone, computer outlets, supplies (e.g., photocopy paper) and convenient work surfaces.

- *file cabinets and other office equipment*
- *space for reference materials:* computer manuals, supply catalogs, phone books, etc.
- *first aid kit, associated safety information and accident logs*

Place close to:

- *supply storage*
- *break room*

3. Office for Director

Even if the director has a separate faculty office, as is the case with most university herbaria, a small office in the herbarium administrative area can be very useful. In addition to providing a place for the director to attend to administrative matters, the office can be used to accommodate sensitive records and confidential discussions.

Equip with:

- *confidential filing cabinet*
- *desk and extra chairs*
- *telephone*
- *networked computer* (optional)
- *confidential fax line* (optional)

4. Office for Curator/Collections Manager

In a large herbarium, a separate office for the curator or collections manager is generally essential, not only to reduce distractions but also to allow for confidential conversations.

Place close to:

- *specimen processing and cataloguing areas*
- *clerical area*
- *areas relevant to any public service function* (e.g., collections, library, reception)

Equip with:

- *desk(s), computer and telephone outlets, bookshelves, file cabinet(s), specimen work surfaces, herbarium case, dissecting microscope*
- *items needed for any non-collections management responsibilities* (e.g., research, teaching)

E. DESIGN CONSIDERATIONS FOR OTHER FACILITIES

1. Research Offices

These offices are designed primarily for staff research on herbarium specimens: therefore some details will be specific to the individual researcher (Fig. 9). In addition, appropriate space should be provided to house research associates, and anyone else needing dedicated space.

FIGURE 9. UC-JEPS research office customized for individual researcher.

Place close to:

- *collections area* (particularly the component that represents the researcher's specialty)
- *library*
- *reception*
- *break room*

Equip with:

- *abundant counter space*
- *herbarium cases*
- *bookshelves*
- *telephone and computer outlets*

Provide access to:

- *sink*

2. Library and Archives

Security is of special concern, since the library generally contains items that can be easily stolen and have resale value. The most valuable items should receive special attention, such as locking them in a glass-fronted cabinet, but without creating unacceptable bottlenecks for authorized users. Visitors without full library privileges might be accommodated by bringing selected materials to a supervised area. UC-JEPS Phanerogamic library contains most of the features listed below.

Place close to:

- *research offices*
- *functions such as public service* which routinely use the references housed in the library.

Equip with:

- *sufficient linear feet of book shelves* to house existing library materials, with room for continued growth.
- *space for reprints, maps, CD-ROMs, video cassettes, microfiche collections, photographic slide collections, card files, various archival materials*
- *counter space and outlets for networked computer* and other equipment (e.g., microfiche reader)
- *at least one good-size table*
- *surfaces on which a book can be opened within convenient access of all shelves*
- *comfortable reading chair* (optional)
- *area to display new acquisitions*
- *designated drop-off point for contributed materials* (e.g., staff reprints)

Provide access to:

- *photocopy facilities*

3. Break/Conference Room

Even the smallest herbarium generally has a corner where staff can take a break, fix a cup of coffee or tea, and chat with visitors and coworkers. This is also the logical place for a bulletin board for general announcements. At UC-JEPS, the break room (also called common room) doubles as a conference room suitable for staff meetings, seminars, interviews, and confidential discussions. So that these uses are as compatible as possible, the food preparation area is in an adjoining kitchenette with an alternate entrance. Remember, however, that unattended heating units used for food preparation represent a potential fire hazard, and eating areas in which food is discarded or stored can attract and provide a breeding ground for pests. A *Stegobium* outbreak in the UC-JEPS herbaria, for example, was finally traced to an abandoned packet of hot chocolate.

Separate from:

- *collections storage and processing areas*
- *library*

If used for seminars, place close to entrance and equip with:

- *shades on any windows*
- *screen* (or at least a blank wall)
- *outlet and shelf* for projection equipment
- *blackboard (or equivalent)*

4. Flexible Space

If possible, allow for at least one flexible space that can be used for a variety of activities, including those unforeseen at the time of planning. Plans for UC-JEPS incorporated three such general purpose rooms, two of which were subsequently used to house an NSF-funded computerization project and an endowed Center for Phycological Documentation neither of which existed at the time the floor plan was finalized.

Equip with:

- *abundant electrical, computer, and telephone outlets*

5. Other Facilities

While the focus of this compilation is on collections-related activities, it is also important to consider the needs of the staff and visitors involved. These might include:

- *a private place*, preferably with a phone, where confidential conversations can be held (e.g., interviews, disciplinary actions).
- *a place to rest:* keep in mind that older researchers and others may require an occasional rest break.

- *a place where receptions, parties and celebrations could most appropriately be held.*
- *specialized affiliated needs* (e.g., botanical artist, laboratories, greenhouse): details beyond the scope of this paper.

CONCLUSION

The preceding synopsis summarizes my personal experience for the benefit of anyone else faced with the challenge and opportunity of designing herbaria. I suggest that one should take advantage of any chance to talk to other people who have had similar experiences. It is important to pay attention to both successes and failures in design as they are equally informative. One should always keep in mind that no design can perfectly address all aspects of conservation and functionality, especially in an era of diminishing resources. The need to compromise becomes inevitable. The goal, nevertheless, is to attain the unique solution that most satisfactorily addresses the specific constraints and needs of each herbarium, with full awareness of how that can best be accomplished.

Acknowledgments

I would like to acknowledge and thank the staffs of UC-JEPS and TEX-LL who assisted with the various herbarium plans, as well as the many architects and staff at other herbaria who shared their own expertise. Art Slater, pest control specialist at UC-Berkeley, provided invaluable advice on pest control matters. Special thanks to Tony Morosco and Irene Acosta for assistance with preparing figures. The efforts of Deb Metsger and Jenny Bull, well beyond the call of editorial responsibility, are sincerely appreciated. Funds for compactor carriages, additional cases, and technical assistance for the various moves of the UC-JEPS Herbaria were provided by National Science Foundation Grants BSR-84-17804, BSR-89-07439 and DEB-91-319871.

Literature Cited

Ertter, B. 1999. Moving herbarbia: A case study at Berkeley. Pp. 181-186 in D. A. Metsger and S. C. Byers, editors. *Managing the Modern Herbarium.*

Society for the Preservation of Natural History Collections, Washington, DC, xxii+384 pp.

HALL, A. V. 1988. Pest control in herbaria. Taxon 37:885-907.

LINNIE, M. J. 1990. Pest control in museums — the use of chemicals and associated health problems. Museum Management and Curatorship 9:419-423.

LULL, W.P. AND MOORE, B. 1999. Herbarium building design and environmental systems. Pp. 105-118 in Metsger, D. M. and S. C. Byers, editors. *Managing the Modern Herbarium*. Society for the Preservation of Natural History Collections, Washington, DC, xxii+384 pp.

RABELER, R. K. 1999. The installation of compactors in herbaria: factors to consider. Pp. 146-164 in D. A. Metsger and S.C. Byers, editors. *Managing the Modern Herbarium*. Society for the Preservation of Natural History Collections, Washington, DC, xxii+384 pp.

STRANG, T. J. K. 1999. A healthy dose of the past: a future direction in herbarium pest control? Pp. 59-80 in Metsger, D. M. and S. C. Byers, editors. *Managing the Modern Herbarium*. Society for the Preservation of Natural History Collections, Washington, DC, xxii+384 pp.

STRANG, T. J. K. 1999. Sensitivity of seeds in herbarium collections to storage conditions and implications for thermal insect pest control methods. Pp. 81-102 in Metsger, D. M. and S. C. Byers, editors. *Managing the Modern Herbarium*. Society for the Preservation of Natural History Collections, Washington, DC, xxii+384 pp.

Chapter 7

The Installation of Compactors in Herbaria: Factors to Consider

Richard K. Rabeler

Abstract.—The major factors which must be considered before a high-density mobile storage system, or compactor, can be installed in a herbarium include: the floor loading capacity; efficiency of the system design; the type of storage unit; rail designs; flooring; mode of operation; aesthetics; other associated improvements required of the physical facilities; and staging the installation. Each factor is discussed, both in general terms and as applied to the recent installation at the University of Michigan. Careful planning, including a thorough assessment of the herbarium early in the process, is emphasized.

Improvement and diversification of collections are major activities of modern herbaria. While such activities are signs of health and vigor, problems quickly arise when cases are filled and there is no more room for specimens. When that point is reached, expansion of the collection must be considered. If additional space is not available to add more cases, the only option is to use the existing space more effectively. One popular way of maximizing the use of available floor space involves installation of high-density mobile shelving systems, or compactors.

Touw and Kores (1984) presented an extensive summary and explanation of the factors that should be considered prior to the actual installation of a compactor system, emphasizing that careful planning is absolutely essential. I have participated in two compactor installations at the herbaria at Michigan State University (MSC) and the University of Michigan (MICH) and have used a number of systems at other herbaria. This combined experience has made me keenly aware of a number of the factors mentioned by Touw and Kores. I have drawn on my experience to discuss these factors further. The figures and specifications cited are based on the compactor at MICH and will differ from those of other design or manufacture. Factors are presented in the order we addressed them; the physical requirements and constraints were of necessity dealt with first. However, regardless of the sequence in which they occur, current use and organization of the collections and requirements for access to them should also be given careful consideration throughout the process.

FACTORS TO CONSIDER

Floor Loading

The floor loading, or weight that a floor can safely support, of the area that will house a compactor is often a critical factor, especially in situations where the proposed installation involves an existing, often older building. Floor loading consists not only of the cases but all of the components of the compactor: the rails, new flooring or decking, carriage components (frames, drive shafts, motors, end panels, etc.) and any attached lighting. While this is not likely to be a problem in a basement (e.g., MSC), it could be for herbaria located on upper floors. Additional I-beams were installed under the floor directly beneath the rails at the Vascular Plant Herbarium of the Biosystematics Research Center in Ottawa (DAO) to increase the load capacity (Barr et al., 1987). The building that MICH occupies was originally a maintenance facility which had been constructed to house heavy equipment so the second floor did not have to be strengthened prior to compactor installation. Because some building designs may not allow floor reinforcement, floor loading should be determined *very* early in the planning process.

Efficiency of System Design

Since the purpose of installing a compactor is to maximize the space gained for collection storage, careful thought should be given to the design of the compactor, including orientation of the rows, ready access to specimens, acceptable aisle width, and compliance with building codes.

The orientation of the compactor within the space will have an impact on both the amount of space gained and ready access to the specimens. The orientation of the existing fixed rows of cases may not be the most efficient arrangement and so other arrangements should be considered. Physical obstructions, especially support columns, may preclude retention of the existing arrangement; fixed rows of cases can easily be placed around a column, a mobile row cannot. For instance, were compactors to be installed on the west side of the collection room at MICH, five vertical I-beams would prevent aligning the compactor carriages in the same configuration as the fixed rows currently occupying that space (Fig. 1).

The orientation of the rows and accessibility to the specimens are critical questions, especially for collections where there are many users (Touw and Kores, 1984). Fixed rows of cases have the advantage that all of the specimens are always accessible by simply opening a door. In compactor systems, ready access depends on the design: in open-shelving systems the entire module may be closed when not in use, while in systems using cases, outward-facing rows on one or both ends of the modules will always be accessible. Access within the module depends on the design and configuration of the rows. For instance, a compactor designed for a rectangular space could either consist of relatively few long rows, or, if rotated 90 degrees, a greater number of shorter rows. In my experience, the latter is more common.

The width of an open aisle can be designed into the system based on the preference of the user. We aimed for a 45 in. aisle at MICH, similar to that which existed before compactors. The aisle is wide enough to allow us to access the specimens easily and turn one of the filing carts sideways in the aisle.

The passage of the Americans with Disabilities Act (ADA) in

FIGURE 1. The University of Michigan Herbarium. Use of space before renovation.

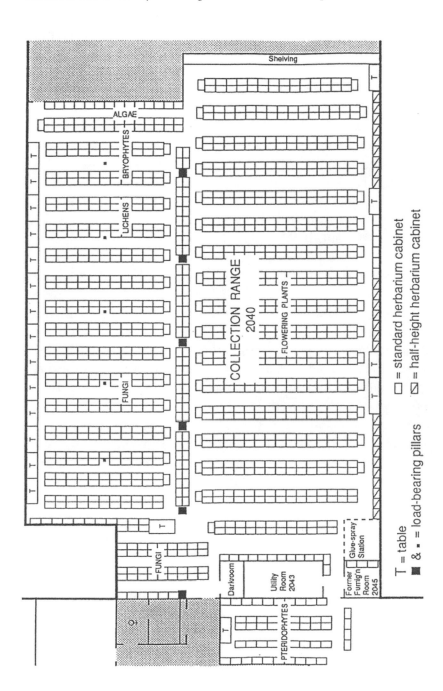

1990 has had a significant impact on building codes throughout the United States. Although specific standards for mobile shelving systems are not given in the ADA codes, general ADA guidelines for aisle width, ramp angles, height of control apparatus and signage, and force required to operate the system should be considered as part of the design (Muth, 1993).

Type of Storage Unit

The decision whether to mount conventional cases onto moveable platforms (i.e., carriages, or "case-and-trays" as described by Touw and Kores, 1984), or to use banks of open shelving involves several factors. These include the total possible space gain, available vertical space, total system weight, condition of existing cases, and finances. Some herbaria have opted to use open shelving (e.g., Missouri Botanical Garden, MO; Academy of Natural Sciences at Philadelphia, PH; and part of the herbaria at Harvard University, GH) or "open-faced cabinets" (DAO; Barr et al., 1987) rather than mounting conventional cases onto carriages. The open-faced cabinets installed at DAO are taller than the standard cases (17 to 20 pigeonholes in height rather than 13) and weigh less than standard cases. This allowed DAO to significantly increase their storage capacity by taking advantage of the approximately 10 ft. high ceilings without overloading the floor (Barr et al., 1987).

While Touw and Kores (1984) noted only one example of a "case-and-tray" installation (the herbarium at the Bishop Museum, BISH), many recent installations have followed that course (MICH; MSC; New York Botanical Garden, NY; University of Georgia, GA; Rancho Santa Ana Botanic Garden, RSA; University of California at Berkeley, UC-JEPS; and University of Wisconsin, WIS; among others). This is an option to consider if an institution already has a large number of cases which are too good (and too expensive) to replace. Carriages can be custom built to match the needs of the installation; MSC has two lengths of carriages, one for rows of Lane cases, and one to accommodate five rows of older cases which are 1 in. wider and 1¾ in. deeper than the Lane cases. Carriages at MICH are 39 ft. 6⅞ in. long, accommodating 16 single (or 8 double and 1 single) Interior cases, with spacers added to other rows holding a mixture of Lane, Sheean, and Interior cases in varying combinations.

Open shelving is preferred by some because of the perceived space saved by the lack of protruding door handles. I would argue, however, that the space gain is minimal and must be balanced against the cost of case replacement. Either eliminating the doors and associated frame protrusion or using new cases with inset handles would provide a gain of about 1 in. per row. However, carriage bumpers and some safety features often require up to 2 in. clearance between adjacent rows. By arranging cases so that the handles are offset in opposing rows, the clearance between rows can be reduced to about the same amount, accommodating the bumpers at the same time (Fig. 2). At MICH, the space saved by offsetting the handles allowed us to add one carriage to the system.

Ease of pest management must also be considered when deciding on the type of storage unit. Despite the advances that have been made in sealing open-shelving systems, such as compressible polyvinyl chloride gaskets used on the DAO units or the solid gaskets at GH (Barr et al., 1987), it is still extremely difficult to localize and control pest infestations in an open system. It remains much easier to localize any pest infestations within the confines of a single conventional case.

Rail Design and Flooring

Two basic rail designs are currently available. One type uses a rail which is virtually flush with the floor with some or all of the rails having a central groove for flanged drive wheels. The other major design uses a rail which is also virtually flush with the floor but includes a channel approximately 1 in. wide on each side of the rail in which the guide wheels that are attached to the non-flanged drive wheels are located. The channels of the latter system must be kept free of debris as they are great resting places for fragments which could drop from the specimen sheets or for dirt tracked in on shoes.

Rails can either be mounted into, or on top of, the floor. While the former is sometimes preferred in new construction, this method can be prohibitively expensive in retrofit installations and

FIGURE 2. The University of Michigan Herbarium. An open aisle.

could be more difficult to repair if differential expansion or contraction of the floor should cause the rails to shift. When the rails
are laid on top of the existing floor, a floor deck is constructed between the rails and ramps are placed along the sides of the outer-
most rails. Since this decking often consists of plywood, the
choice of a finished surface which can withstand some flexing, especially in the case of safety floor panels, must also be made. Carpet (installed at MSC) or a heavy rubber tile (installed at MICH)
are both good choices; vinyl tile is less flexible and is more subject
to cracking or chipping, especially along the edges of guide wheel
channels.

Lighting

Any lighting in the collection area which only illuminates existing fixed rows will need to be augmented or replaced. I have
seen two arrangements that effectively light compactor rows.
Fixed rows of lights arranged perpendicular to the compactor (installed at MSC) work well, but are inefficient because all the lights
in the system must be on even if only one aisle is in use. Lights
mounted on top of the cases which come on only when the aisle is
opened (installed at MICH and part of MO), while relatively expensive to install, should have lower overall energy costs. The decision to have case-mounted lighting is independent of whether
the system is electrically or manually operated, however the cost
for the manual system would be significantly higher. Individually
switched rows of fixed lighting installed where the aisles are calculated to open would have the same effect as case-mounted lighting
but is more difficult to design.

Electrical versus Manual Operation

A well-designed manual system should allow the user to move
a row of the compactor without a great deal of physical strength or
a large number of turns of the crank. There are installations which
violate this assumption, most often requiring *lots* of cranking. The
length and weight of the row being moved can be a critical factor
in this decision, especially in larger systems. The MICH system,
consisting of rows almost 40 ft. in length and the weight of 32 to

34 cases, was near, or in some cases beyond, the limits of available manual systems.

The obvious advantage of an electrical system is that the row moves at the touch of a button, a clear choice in the minds of ADA advocates. Modern electrical systems (such as Spacesaver's Soft-start/Soft-stop™) may provide more protection to specimens than manual systems as they are designed to reduce the motor speed when starting or stopping a moving row. This reduces the jarring motion which may loosen seeds and other fragments from the herbarium sheets when adjacent rows come in contact. However, if the power available to the building is not dependable, the electrical system could be more problematic than the convenience which it should offer. The MICH system has a battery pack which can be used to move rows in case of a power failure.

Safety Features

The ability to move the rows brings not only convenience to users but also safety concerns. How will a user, the cart he/she may be using and any open case doors be protected from possible damage from movement of another row? Several devices are now available which address one or more of these concerns, depending on the manufacturer. Manual systems may have a manual lock which the user engages to lock the desired row open (as at MSC, UC-JEPS), or opening a row may engage a passive lock (as at NY). Depending on the desired level of safety (and available funding), electrical systems can be equipped with a variety of devices: safety floors detect as little as 25 lbs. of weight in an aisle by one of the microswitches in the floor panels, preventing other rows in that module from being opened; pressure-sensing bars stop movement if one pushes against them; and infrared sweeps cause the system to sense the obstruction and stop any motion when the beam, which is projected from one end of the row to the other, is interrupted.

Installations at institutions located in areas which are, or could be, seismically active may have to include anti-tip devices or construction modifications (e.g., bolting all of the cases on a carriage together at UC-JEPS) to comply with state and/or local safety codes.

Aesthetics

While not critical to the operation of the compactor, the colors chosen for the compactor end panels, frames, floor coverings, etc. do have a major impact on the appearance of the system, both to the herbarium staff who use it daily and to visitors. If the installation is part of a renovation project, this could be the only chance to convert a drab area into a bright, more cheerful working environment.

A scheme including brightly-colored end panels and neutral floor colors works well to highlight the compactor. The combination of orange end panels, gray carriage frames, and tan carpeting was used at MSC while a scheme of bright red end panels, black carriage frames, and tan rubber tile on the safety floor was chosen at MICH (Figs. 2, 3).

Other Associated Improvements to the Physical Facilities

If the collection area is not air-conditioned, it might be appropriate to consider this improvement prior to installation of the compactor. While air conditioning will provide a more stable temperature for both the collection and workers, humidity control is an equally important factor. Moisture levels below approximately 40% relative humidity will hinder larval growth of common herbarium pests (Hall, 1988), potentially eliminating the need for chemical pesticides in the collection. Additional information on conditioning the collection environment is covered by Lull and Moore (this volume).

Installation of temperature and humidity controls can be a significant project in itself, depending on the condition of the physical facilities. Prior to the installation of the compactor at MICH, air-conditioning, with humidity control provided by a desiccant dehumidifier with a gas-fired reactivation heater, was installed. In order to effectively create a vapor barrier between the conditioned space (the collection room) and the outside world, many other renovations were made: insulating all outside walls; sealing all floor, wall, and ceiling penetrations; and, painting the entire room with an epoxy paint. Fortunately, previous building renovations

FIGURE 3. The University of Michigan Herbarium. East side of compactor.

provided an insulated roof and thermopane windows, both of which are necessary components in the conditioning equation.

The top of a row of herbarium cases is often a tempting place for storage of everything from pressing cardboards to boxes of specimens. This temptation, however, might be in violation of local fire codes. When the rows are moveable such storage also becomes a safety concern and case-mounted lighting removes even the temptation.

Staging the Project

Having participated in two compactor installations I cannot emphasize the importance of staging enough. Transforming the engineering diagrams for compactors into reality can be a very challenging proposition. While details may vary widely among installations, several common concerns are worthy of further discussion.

Compactor installation begins with laying the rails; therefore, all or a major portion of the space which the compactor will occupy must be cleared. If all the cases can be moved out of the area, such as was done for the MSC project, all of the rail can be laid easily. In the case of systems with carriages aligned parallel to the shorter side of a rectangular room (Fig. 4), the rail can be laid in a continuous fashion for the entire system. This is an important advantage since it allows all (or all but the end rows, depending on design) of the carriages to be moveable. The decision to fix rows (if any) is not determined by rail breaks.

If the cases cannot be moved out of the installation area, the compactor will probably have to be built in segments or modules. Existing fixed rows of cases would have to be moved and/or compressed to clear floor space and then, in turn, be loaded onto completed carriages to clear floor space for construction of the next module. The MICH compactor was built in four modules, requiring clear floor space of about 40 x 40 ft. to allow for proper sighting and leveling of the tracks for each module. Although the four modules are adjacent and appear to form a continuous bank of carriages (Figs. 3, 4), the rails are not continuous between all modules. Each module was leveled independently with a fixed row

FIGURE 4. The University of Michigan Herbarium. Use of space after renovation.

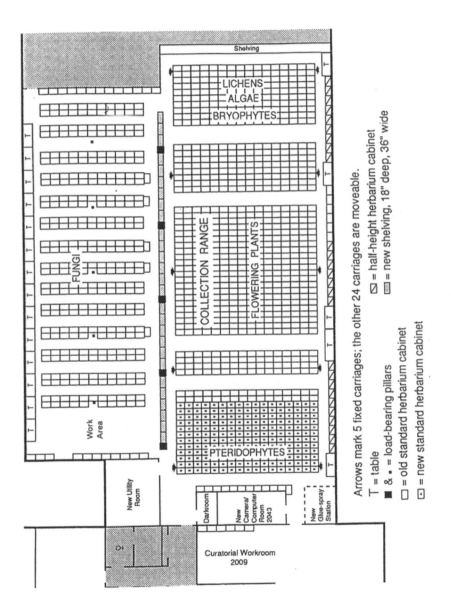

placed at the rail break between the modules. While this imposes three internal fixed rows, it may have been a blessing in disguise; without a difference in elevation between two of the modules, some of the case-mounted lighting would not have cleared the overhead roof support beams (Fig. 3)!

While moving the collection to allow compactor installation may be a major project, the moving of it onto the compactor can be a much larger one if both moves are not carefully planned. I recommend a thorough inspection of the herbarium early in the planning process. If the original cases are to be retained, are they all identical or are they different sizes? If different, how many are there of each size? Do all of the doors open from the same side? Are there any cases which must be replaced (because they are broken, too small or large, etc.)? If one or more of these situations apply (all did at MICH!), constructing a database of the specifications of the herbarium cases is a useful means of summarizing the state of the existing physical herbarium and what must be dealt with when the system is reloaded. The database should include not only contents of the cases but also case number, case height and width, handle position, and location.

If cases are retained, the loading process can be simplified if at least a tentative loading order is known before the cases are cleared from the room. This should allow placing them so they can be accessed when needed while loading the carriages. If all the cases are identical, they can be loaded on the carriages in the same order as they were in fixed rows. When the cases are different and must be loaded in a different order, cases may have to be retrieved from widely scattered places throughout the collection. As a result the collection order will probably be scrambled. A map of the loaded order will be essential both to find items in the collection and, ultimately, to rearrange the collection.

A major goal of installing a compactor is to permit redistribution of the collection to eliminate crowding and uniformly distribute (or, more accurately, correctly apportion) the space available for future expansion of the collection. This redistribution should be carefully planned. Several questions should be considered as the plan is developed. If additional cases or shelving are to be added, will they all be placed at one end of the collection or interspersed within it? If placed at one end, how much of the collection should be redistributed to take advantage of the space? If particular areas of the collection are growing faster than others, how much extra space should be given to these areas?

Space should also be left for specimens which are on loan to other researchers. While some institutions installing compactors have asked researchers to return all loans before the installation begins, I have found it far easier to take a printout of all of the outstanding loans to the collection, verify that each loan was noted by a drop tag, and then estimate how many pigeon-holes each loan occupied.

Besides the obvious financial considerations for such a project, sufficient labor must be available. Who will move the cases to clear the room? Although the cost for labor to load cases onto the carriages of the new compactor will probably be included in the cost estimate of the compactor installation, transferring specimens from cases to open shelving most likely would not. If the order of the collection has been scrambled, the previous order should be recreated before expansion begins, at least in larger herbaria. While doing both at once sounds economical, I suggest that it would be far too complex. Redistribution of the collection will be tedious and labor intensive. However, if care is not taken at this stage, errors which may not be found for many years, could result.

The University of Michigan Herbarium Installation

While I have referred to some facets of the MICH installation in previous paragraphs, presenting a summary of the results of the project seems appropriate. The University of Michigan Herbarium, with about 1.7 million specimens, is, according to figures given in Holmgren et al. (1990), the second largest state university herbarium and the seventh largest collection in the Western Hemisphere. The intent of the compactor installation was to increase the storage capacity of one half of the collection room, the area where the largest portion of the collection, 1.1 million vascular plant specimens, was housed. The original floor plan of the area (Fig. 1) shows rows of standard herbarium cases and fixed aisles which nearly fill the room. The number of cases which could be added without compromising either aisles or visitor work space was fewer than ten. The decision to install a compactor system using existing cases on carriages rather than converting to a compactor system with open shelving was based on two major considerations: the number of cases already at MICH and the investment they represent; and, the height limitation within the room due to overhead roof support beams (the lighting clears these beams by less than 2 in.).

The new floor plan (Fig. 4) illustrates the placement of the Spacesaver S/4 compactor (Spacesaver Corp., Fort Atkinson, Wisconsin) that was installed (Fig. 3). The unit occupies 4500 sq. ft. and is 113 ft. long, with individual rows measuring about 40 ft. including carriage length, end panels, and supports. The compactor was built in four modules, each with one moveable aisle. Each of the 29 carriages (5 fixed and 24 mobile) holds 2 rows of 16 or 17 cases and weighs about 17,000 lbs. The approximate weight of the entire system is 508,000 lbs., an imposed load of 112 lbs./sq. ft. under the system. Dividing this load by the square footage of the room (6000 sq. ft.) gives a superimposed live load across the structural system of about 85 lbs./sq. ft. This is under the calculated 100+ lbs. (including a safety margin of 60%) live load capacity of the floor in the room. This is probably representative of calculated loads for standard herbarium installations.

Each aisle is illuminated by nine 2-tube fluorescent fixtures attached to the top of each mobile row of cases (Fig. 2). Only the central 34 ins. of each aisle is not directly under a fluorescent light since this space over most aisles is occupied by an electrical conduit, the pantograph, which must fold when the aisle is closed. The compactor system is electrically operated and includes both a safety floor to sense aisle occupancy with weight sensors mounted on the edges of the floor panels in the guide wheel channels, and an infrared floor sweep to sense objects near floor level. Installation of the compactor was completed in about eight weeks.

The impact of the compactor installation is best seen by comparing Figures 1 and 4. Before the compactor was installed, the space housed 536 cases in 34 rows with 16 fixed aisles (Fig. 1). This same space with the compactor holds 940 cases arranged in 58 rows with 4 moveable aisles (Fig. 4), a net gain of 404 cases.

This represents a 75% increase in the number of storage cases. Since 200 of those 404 cases came from other areas within the collection room, we were able to use the space formerly occupied by those cases for a new common equipment room, an expansion of our curatorial workroom, and a new visitor area. As well, it was possible to install 500 linear ft. of heavy duty storage shelving to hold many of the items formerly stored on top of the herbarium cases.

Staging the MICH installation was complex. The existing cases at MICH were from four different manufacturers (Interior Steel, Lane, Sheean, and Steelcase), in five widths and had doors opening from both the right and the left. The Steelcase cases, which were 3 in. taller than either Lane or Interior cases, were too tall given the height restrictions of the overhead beams. Critical data about each case was entered into a database and the entire collection was manually mapped as mentioned earlier. After the length of the carriage had been determined, we used information from the inventory of cases to determine the optimal combination of existing cases which would fit into that length. This step led to eliminating the old, narrow (27 in. width) Lane cases from the compactor. To account for handle positions as well as placement of the three remaining sizes of cases, a map of the loaded compactor was prepared before any loading was begun. Similarly, to account for the location of all cases at all times, maps were updated as cases were either cleared from one section of the room or loaded onto the compactor. When the loading of the compactor was complete, the vascular plant collection was completely scrambled; one case was in the correct place according to the pre-expansion arrangement. Color-coded maps allowed two staff members to restore the collection to its original order over the course of eight weeks.

Three factors were considered when devising the formula for redistributing the collection. The goal was to have all pigeonholes no more than three-quarters full; to provide one empty pigeonhole per case, or two in cases on the ends of the rows (where an increased reach was necessary to access those pigeonholes in cases over the ramped floor panels); and, to accommodate specimens which were on loan to other institutions. We analyzed the accession records for the past seven years to locate the fastest growing areas of the collection. Not surprisingly, they corresponded to areas of active faculty research. Additional space in these areas was allocated in proportion to growth over that period. The redistribution formula was reviewed periodically to ensure that we would not exceed the total number of pigeonholes available. This task was completed by one conscientious staff member in approximately 18 weeks.

CONCLUSION

As evident from the preceding paragraphs, making the decision to install a compactor is only the first in a series of decisions that must be made. Fortunately, anyone making that decision is not alone and can ask others in the community about the process. I encourage not only asking but, if at all possible, visiting other sites. The best way to see how different designs, controls, and safety systems work is to visit sites that have them. I feel that the MICH installation benefitted from visits that our staff made to other herbaria and libraries.

ACKNOWLEDGMENTS

I would like to thank Anton A. Reznicek and Beverly Walters for their assistance, both mental and physical, in bringing the installation to a successful completion; Anton A. Reznicek for reviewing early drafts of this paper; William R. Anderson for preparing the floor plans shown in Figures 1 and 4 as well as conducting most of the early planning and fundraising for the project; James Solomon (MO) and Gisele Mitrow (DAO) for providing literature and/or citations; Barbara Ertter (UC-JEPS) for information on the UC-JEPS installation; and Jacquelyn Kallunki (NY) and Christine Niezgoda (F) for their reviews of this manuscript. Approximately $1 million for the MICH project was provided by the National Science Foundation (Long Term Projects in Environmental Biology Program) and The University of Michigan (College of Literature, Science and the Arts and the Office of the Vice President for Research).

LITERATURE CITED

BARR, D. J. S., W. J. CODY, AND J. A. PARMELEE. 1987. State-of-the-art herbarium compactor systems for DAO and DAOM. Taxon 36:413-421.

HALL, A. V. 1988. Pest control in herbaria. Taxon 37:885-907.

HOLMGREN, P. K., N. H. HOLMGREN, AND L. C. BARNETT. 1990. Index Herbariorum. Part I: The herbaria of the world. Regnum Veg. 120:x + 1-693.

LULL, W.P. AND MOORE, B. 1999. Herbarium building design and environmental systems. Pp. 105–118 in Metsger, D. M. and S. C. Byers, editors. *Managing the Modern Herbarium.* Society for the Preservation of Natural History Collections, Washington, DC, xxii+384 pp.

MUTH, J. 1993. *Achieving ADA compliance with mobile storage.* Spacesaver Corp., Fort Atkinson, Wisconsin.

TOUW, M. AND P. KORES. 1984. Compactorization in herbaria: planning factors and four case studies. Taxon 33:276-287.

CHAPTER 8

The Brooklyn Botanic Garden Herbarium: A Case Study in Modern Herbarium Design

KERRY BARRINGER

Abstract.—The Brooklyn Botanic Garden Herbarium was moved into a new facility in November 1991. The facility was designed with features to help control insect infestations and atmospheric pollutants; prevent damage by fire and water; and provide adequate but safe light levels. The new facility was evaluated after almost five years. The results indicated that herbarium cases provide the first line of defense in protecting specimens and so should be sturdy, well-sealed and light colored. In addition, insect infestations are effectively controlled by isolating the collection and keeping the herbarium temperatures below 62°F (18°C). Gaseous filtration is effective in removing both insecticide residues and particulates from the collection but is not always 100% effective in removing atmospheric pollutants from areas with excess contamination from external sources. Inexpensive, indirect fluorescent light fixtures combined with light-colored cabinet finishes and a white suspended ceiling provide an appropriate light source for the herbarium. These adaptations, along with a more accessible organization of the specimens, have transformed this excellent herbarium from a poorly-curated collection to a safe, accessible and useful one.

In November 1991, the Brooklyn Botanic Garden (the Garden) moved its herbarium (BKL) into a new facility designed to provide an appropriate conservation environment for the herbarium specimens. The new herbarium was part of the Garden's capital renovation project that also saw the renovation of the Administration Building and construction of new conservatories. The former herbarium facility had been neglected for many years. It was overcrowded and had become completely inadequate for its new role as a research tool (Lull, 1991). To accommodate the growth of the research departments, the Garden purchased an eighty-year-old warehouse and office building adjacent to the Garden grounds and prepared to renovate the building to house the library, herbarium, laboratories, and offices.

Prior to designing the new facility, the herbarium staff identified a set of actual and potential problems that needed to be addressed in this collection along with desired attributes for a new facility. The list included damage to the collections from insect infestations; contamination by atmospheric pollutants; the need to control temperature and humidity; the need to guard against disastrous damage by flood or fire; the provision of illumination that was both adequate for specimen observation and non-damaging to the specimens; and, an appropriate organization of the collections to make them accessible beyond the taxonomic community. In order to address many of these issues properly, the Garden received a Conservation Grant from the Institute of Museum Services (IMS) to have conservation environment specialist William P. Lull examine the extant facilities and make recommendations concerning the design of the new herbarium (Lull, 1991). In 1993, the Garden received a second IMS Conservation Grant to bring in conservator Catherine Hawks to conduct a conservation assessment of the collections and a preliminary evaluation of the herbarium (Hawks, 1993). The problems identified in the old facility were extreme, but the root causes of the problems were the same as those affecting all herbaria. Working with these conservation specialists, we were able to redesign both the herbarium facilities and the procedures used in them to significantly improve our ability to maintain and care for the collection.

Though the renovation is long since completed, we are constantly re-evaluating our ideas and procedures. During these first five years of operation, the facility has already been subjected to a series of unusually hot and humid summers and two very cold winters. Consequently, we have had to rethink and adapt our tools and techniques according to their success or failure under differing conditions and as our understanding of preventive conservation methods improves with experience. This paper summarizes the results to date.

CONTROLLING INSECT PESTS

The main threat and the cause of most past damage to specimens in BKL was infestation by insects. The main culprits were the cigarette beetle, *Lasioderma serricorne*, and, to a lesser extent, the drugstore beetle, *Stegobium paniceum*. These infestations were the result of the interaction of many factors. First, the specimens were stored in cupboards of enclosed steel shelving. These cupboards were built in long rows with no internal barriers between shelves in the row. The doors of the cupboards were gasketed with felt and were difficult to close tightly due to settling of both the floor and the units over time. The cupboards were not an effective barrier to insect movement but rather encouraged it.

Second, the building was not air-conditioned. The room was cooled in the summer by opening windows and turning on fans. This allowed pests free access to the collections. During the summer, a small cafe for Garden visitors was located on a terrace next to the herbarium. The food and garbage from the cafe attracted roaches, both the German cockroach, *Blattea germanica*, and the brown-banded cockroach, *Supella supellectilum*. The roaches, as well as the yellow-jacket wasp, *Vespula vulgaris*, were frequent visitors to the herbarium. Third, the room that held the herbarium cases in the old facility also served as work space for the curators, with desks and work tables squeezed in among the cases. Thus the collection could not be isolated either to fumigate or to prevent the introduction of pests. Plants from the field and from the Garden were handled and studied alongside herbarium specimens. Curators had food and drinks at their desks. Proper cleaning was not always possible because the custodians could not disturb the work surfaces.

We did not want to use chemical pesticides to control the

insect infestations as all of the commonly-used pesticides could potentially damage the specimens, the environment, and anyone using the collection (Peltz and Rossol, 1983). Rather, preventing the infestations became paramount. The central concept to prevention is to isolate the collection in an environment that discourages insect infestation. In our case three lines of defense were chosen: the provision of well-sealed herbarium cases; maintenance of ambient temperatures and humidity that are low enough to discourage insect activity; and isolation of the herbarium from offices and workrooms.

Herbarium Cases

The herbarium cases selected for the new herbarium are made of heavy-gauge steel, completely seam-welded to eliminate gaps. The doors of the cases are gasketed either with silicone or with ethylene vinyl acetate closed-cell foam to ensure a well-fitting seal. The cases are finished in a white powdercoat inside and out which makes it easier to see the insects. Cases fitted with recessed handles were selected both because they are less likely to snag clothing and because they will be easier to mount on compactor sleds in the future. We purchased the standard herbarium cases available from Interior Steel Equipment Company and have been very pleased with both the product and the service provided by them. Since the original purchase, oversized cases have been purchased to accommodate loans from institutions with larger-sized sheets. The oversized cases take up little additional space while greatly increasing options for loan requests. In addition, it is easier to file and remove sheets because the folders can be held on the sides as well as from below.

Temperature Control

Ambient room temperatures between 60° and 62°F (15° and 18°C) were targeted as these temperatures are known to limit breeding and development of the herbarium insect pests noted above (Strang, 1992a, 1992b). Maintenance of these temperatures required the installation of a separate heating, ventilation, and air conditioning (HVAC) system for the collection room capable of operating with minimal maintenance 24 hours-a-day, 365 days-a-year.

These low temperatures are uncomfortable to work in for long periods of time. Thus, a review of patterns of herbarium use was conducted to determine who would be using the collection for what length of time. This facilitated the assignment of alternative work areas. The herbarium is used primarily by Garden staff. The research staff study specimens in their offices which are now located outside the herbarium. Visitors and other staff usually use a workroom just outside the herbarium, where specimens can be examined in comfort. Anyone filing specimens or quickly examining a large number of specimens is required to work in the herbarium and so has to adjust to the conditions there. People have complained about the cool temperatures in the herbarium. Our primary solution has been to provide sweatshirts!

Treatment Procedures for Incoming Material

With the collection properly isolated, housed, and environmentally controlled, specimen handling procedures were revised, so that, as much as possible, specimens are kept in the herbarium unless they are being actively used. All incoming collections are processed well away from the herbarium: loans and gifts in a prep room located near shipping and receiving; recent field collections in a dedicated pressing room adjacent to the garage. Prior to entering the herbarium all specimens, whether new, or extant specimens that have been removed for loan or study, are frozen at −20°F (−28°C; Florian, 1986, 1990) for at least five days to kill all stages of insects.

Two upright 15 cu. ft. (0.43 cu m) freezers, an older Kenmore model and a Magic Chef are used for pest control. Both units can maintain a steady temperature of −20°F (−27°C) but differ in their specimen capacity. The Kenmore can hold more than a case of specimens, while the Magic Chef had to be modified in order to hold about a case of specimens. The modification involved removing the shelves on the door and replacing them with a piece of sheet aluminum and extra insulation to provide an additional 3 to 4 in. (10 cm) of depth. Despite the fact that it invalidated the warranties, the modification increased the loading capacity of the unit, making it possible to load the herbarium sheets lengthwise, two stacks per shelf.

Freezing has proved to be an effective pest control treatment but both higher temperatures and shorter exposure time in the freezers reduce the effectiveness of the method. For instance, Bridson and Forman (1992) suggested that the loss of cold air from opening and closing upright freezers increases the chill time required to bring all specimens down to an effective killing temperature.

Dealing with Infestation in the Main Collection

Because the main collection was known to be infested prior to the move to the new facility, a procedure was developed to help isolate infested specimens within the herbarium. When the collection was moved from the old facility to the new one, the specimens in each of the twenty-four shelves in a cupboard were wrapped as a single package. The packages of specimens were moved to the new herbarium and filed, still wrapped, in the new cases (Barringer, in prep.). Case by case, the infested specimens were frozen. While the specimens were in the freezer, the case they had occupied was cleaned and vacuumed with a vacuum cleaner designed for office equipment. This vacuum uses the fine-pored filter bags designed for photocopy toner. The filters are not as fine as a HEPA filter, but are still capable of removing particles to about 5 microns in diameter. They were chosen over HEPA filters because they are about half the cost. After the case was cleaned and their freezer treatment completed, the specimens from a single case were unwrapped and filed in the clean cases. The method of filing is discussed in the section on organization.

Unfortunately, this procedure did cause some problems for those using the collection. Because of the limited freezing capacity, many of the specimens remained wrapped long after the move. If a specimen that had not yet been frozen was needed for study, a packet of specimens would have to be unwrapped. Even though the contents were marked on each packet, some people found it difficult to locate a particular specimen. Also, curators had to consider the freezing schedule in planning their work. Occasionally specimens desired for immediate observation were in the freezer. At other times, curators wanted to refile specimens that they had removed "for just a minute" without refreezing them. As BKL is a small herbarium, we have been able to work through these difficulties with a little forethought. These procedures may not, however, be workable in a larger herbarium or a collection with many more visitors. A larger freezing capacity, as well as a tracking mechanism to indicate which specimens are currently unavailable and when they will be refiled, would improve the efficiency of the treatment process.

Evaluation of the Effectiveness of Pest Control Measures

To address the concern that insect damage to the specimens might be continuing despite the new controls, the collection was monitored using pheromone traps with an additional food lure (Gilberg and Brokerhof, 1991; Gilberg and Roach, 1991). By monitoring the traps and some specimens we determined that insect activity was almost completely curtailed when the herbarium temperature was below 62°F (18°C) and the relative humidity (RH) was below 50%. The extent of the insect infestation was estimated on three occasions when the HVAC unit froze up and needed repair. At these times, air from the rest of the building was allowed to circulate through the filters into the herbarium and the temperature rose to between 75° and 80°F (24° and 27°C). The infestation was estimated by examining the pheromone traps during the ten to fourteen days that the HVAC unit was out of service. On the first occasion, during the summer, seven months after the move, we found isolated insect activity only in specimens which had not yet been frozen. On the second occasion, thirteen months after the move, insect activity was evident in only two cases. The first case held specimens of Annonaceae which had been frozen for only 48 hours when we were still following the recommendations of Bridson and Forman (1992). The second case held specimens of cultivated Rosaceae which had not yet been frozen. On the third occasion, twenty-five months after the move, insect activity was found in only one case which contained Aristolochiaceae, another group which had been frozen for only 48 hours. Since then, the temperature in the herbarium has risen on two occasions, the first when the condenser on the HVAC unit was upgraded and the second when the humidification unit was replaced. We did not observe any insect activity in the collection at these times.

Effectiveness of the HVAC Unit at Maintaining Temperature and Humidity Levels

During the first year and a half of occupation, we had a problem maintaining the temperature and RH in the herbarium at the levels necessary to control insects. During Brooklyn's hot, humid summers, the HVAC unit would freeze up if we tried to maintain a temperature below 68°F (20°C). The manufacturer, BioCirc Corporation, evaluated the system and found that the HVAC unit could not handle the necessary load, so the condenser was upgraded. This has greatly improved the performance of the unit but we still have a slow seasonal fluctuation in RH that should be reduced. Also, the underside of the roof is visible above the suspended ceiling of the herbarium. In summer, roof heat radiates downward, heating the herbarium and increasing the load on the unit. Installation of additional insulation and a vapor barrier between the suspended ceiling and the roof is planned, to try to reduce the load on the unit. This structural modification should allow the unit to maintain a more stable temperature and RH during the summer months.

ATMOSPHERIC POLLUTANTS

Herbarium collections can be damaged by many different gaseous and particulate pollutants (Baer and Banks, 1985), some of which, but not all, come from outside the collection (Woods, 1983; Lull, 1991; Lull and Moore, this volume). Sulphur dioxide, oxides of nitrogen, ozone, and particulates including soot, are common urban pollutants that are generated by both industry and transportation and can damage collections. Formaldehyde and other organic solvents released by building materials, as well as residues from insecticides, adhesives, liquid preservatives, and other chemicals used in herbaria, can acidify or degrade specimens and their mounts.

At the outset, BKL stunk. The smells were not just those that normally come from specimens of *Penstemon* (Scrophulariaceae) or species of Valerianaceae, but were also the odors of pesticides which had been used in the collection in the past. At one time, the gardener in charge of pest control on the grounds also took care of the building and the herbarium was routinely sprayed and fumigated. While the record of pesticide use is incomplete, there are records of the more recent use of Vapona (dichlorvos), paradichlorobenzene (PDB) and naphthalene. Some of the older specimens show the characteristic black staining that results from poisoning the specimens with mercuric chloride (Clark, 1986). Metallic mercury vapor is continuously released from these treated sheets and may become a hazard in poorly ventilated rooms (Briggs et al., 1983).

Gaseous and Particulate Filtration

Lull (1991) suggested the addition of special filters to the HVAC system to control both gaseous and particulate pollutants. Particulate filters and an activated charcoal/potassium permanganate gaseous filtration system were included in the HVAC unit. All the air brought in from outside must pass through these filters before it is moved into the collection room. Air from the collection room is recirculated through these filters. This filtration provides excellent control of dust and soot and makes cleaning the collection room much easier. Gaseous filtration has reduced the levels of pollutants far below what they were, but at certain times the chloride ion and NO_2 levels have been higher than desired (Williams, 1989; Hawks, 1993). This is a concern because both these ions are highly reactive and can combine with water vapor to form acids. In the long term, these pollutants are absorbed but over short periods of time, the amount of pollution still seems to exceed the capacity of the filters. It appears likely that this problem stems from the air drawn in from outside the building, but the exact source has not yet been identified. The Garden is located adjacent to some light industry, including a commercial laundry, which may be the source of some of the substances detected. However, chloride ions might also be derived from the breakdown of the residual pesticides on the specimens and mounting sheets. Gaseous monitoring has been increased in an attempt to isolate the source and address the problem.

Use of Appropriate Building Materials to Minimize Contamination

Construction materials used in the collections room and surrounding areas were chosen to minimize offgassing. Lull (1991) recommended outfitting the herbarium with finishes and materials that do not offgas. Specifically, he recommended powdercoat finishes for the herbarium cases instead of paint; linoleum rather than vinyl flooring which contains plasticizers known to offgas; latex-based paints rather than oil-based paints; and finally, metal rather than wood shelving.

The powdercoat finish on the herbarium cases has proved to be superior to paint. All the cases were given white finishes, but the painted cases had a strong solvent smell when purchased and have yellowed in comparison to the powdercoated cases. The manufacturer suggests that the smell and yellowing are due to inadequate curing after the cases were painted. This is apparently common when light-colored paints are used because the painter often does not want to risk discoloring the finish with the high heat needed to fully cure it (Interior Steel Equipment Corporation, pers. comm.)!

The linoleum flooring has proven to be durable and easy to maintain. Since it is uniform, scratches and gouges do not ruin the floor. Even after a water leak lifted the linoleum, we were able to glue it back in place and it continued to look good. Linoleum can be difficult to purchase as it is no longer manufactured in North America, but it is widely used to cover theatre stages as it is preferred by dancers for its resilience and flexibility (Rothzeid, Kaiserman, Timmerman, and Bee, Architects, pers. comm.). We purchased our linoleum from Forbo Industries, Inc., New York.

WATER

Humidity Control

Changes in humidity can cause expansion and contraction of organic materials, especially woody specimens. Differential expansion and contraction can damage specimens, mounting media and paper. Extremely high humidity (above 70% RH) can lead to mold growth and can promote acidification and oxidation of residual compounds on the sheets. Extremely low humidity (below 35% RH) can cause extreme shrinkage and make specimens too brittle to handle. Water can also cause catastrophic damage through leaks and flooding (Clark, 1986).

In the short term, herbarium cases will buffer the effects of variable humidity. However, for long-term control, the humidity in the collections room needs to be regulated. Lull (1991) recommended that a stable humidity in the range of 35 to 50% RH would be appropriate for the dry specimens.

In practice, it has been as difficult to maintain a stable RH in the collections room as it has been to maintain low temperatures. Humidity is more difficult to control at low temperatures. The humidity in the herbarium tends to parallel changes in the outdoor humidity, though in a much smaller range. The causes for this fluctuation may include: the inability of the refrigeration unit in the HVAC system to remove enough water from the air stream; an improper adjustment in the programming for the final stage reheat of the HVAC system that brings the air up to the correct temperature after it has been cooled to remove the water; an overload of the HVAC unit's capacity to treat outside air; and various problems with the roof as mentioned earlier. These factors are being investigated one at a time and corrected as required.

Preventing the Occurrence and Effects of Water Damage

Well-sealed cases should protect specimens from most leaks and floods provided they are not submerged for long periods of time. However, to guard against such incidents it was recommended that the roof either be sealed or replaced, and the existing roof drains installed with drip pans. No piping, except for the sprinklers, was to be allowed over the collections room. A spill cart with barriers, absorbent pillows and a wet/dry vac were to be made available in the building in the event of a flood.

The first heavy rain demonstrated that the roof had not been adequately sealed and that drip pans had not been installed. Ceiling tiles were damaged and linoleum became unglued when water dripped through the ceiling. Thankfully, the cases protected the specimens from water damage. Fortunately, all drips occurred over aisles rather than cabinets and so the exposed fluorescent fixtures on top of the cases (see Lighting below) were unaffected. The roof was under warranty and so it has been resealed and the places where drains pierce the roof have been releaded. It does not leak now, but will need to be replaced in the future. Drip pans will be installed around all drains.

FIRE

The dangers of fire in the herbarium are greatly reduced when steel cases are used, but these cases do not provide adequate thermal insulation to protect specimens during a building fire. Wet collections are especially vulnerable because volatile vapors can build up in areas where wet collections are stored. Alcohol fires form little smoke and so can go undetected until they spread (Lull, 1991). The fire codes of some cities limit the amount of alcohol or other flammable liquids that can be stored in a building without special facilities and special permits. This is not a problem in our collection, but could be for others (Hawks, 1993).

While well-sealed steel cases form the first line of protection against fire, sprinkler systems are also an important defense and are required by building codes in most cities. Lull (1991) recommended flow-control sprinkler heads which, unlike typical sprinkler heads, shut off after the fire is extinguished. These heads are sensitive to heat. Smoke detectors have also been installed throughout the building. The wet collections are kept in a separate room adjacent to the herbarium. The walls of that room are built with two layers of sheet rock and the room is vented separately to decrease the concentration of fumes. Both water and dry chemical fire extinguishers are available throughout the new facility.

Fortunately, the system has never been put to the test in a real fire. However, the building has been inspected by the local fire department and they are pleased with the preventive measures taken.

LIGHT

Many herbaria are plagued by inadequate light levels in parts of their collections which can make working in a collection difficult or impossible. Our goal was to provide good, overall lighting even to the bottom shelves of the herbarium cases, and flexible task lighting over work surfaces.

Sunlight is very damaging to specimens and, if possible, they should never be exposed to it. Thus it is best if the collections room is without windows. Fluorescent lights are also a source of ultraviolet (UV) rays that can damage specimens (Lafontaine and Wood, 1982; Lull, 1992).

For our situation, indirect fluorescent lighting was chosen as the best quality and most cost effective overall source of illumination. We did not place UV filters over the fluorescent bulbs because herbarium specimens are kept in closed cases most of the time and therefore the exposure to UV would be very low.

In the offices the indirect lighting is combined with direct fluorescent or incandescent task lighting over work areas. In the herbarium, 48 in. (1.22 m) fluorescent fixtures are attached to the tops of the herbarium cases using small brackets built onto the cases. The fluorescent fixtures are very inexpensive and the bracket was attached by the case manufacturer at no additional charge. The fixtures are pointed upward so the light reflects off the white suspended ceiling. Since all the cases are white, the maximum amount of light is reflected down to the floor and into the cases.

The lighting in the herbarium has proven to be excellent (Hawks, 1993). Light is bright and specimens are easy to see even in the bottommost shelves of a case. The light is, in fact, so good that color photographs have been taken in the room without supplemental lighting.

ORGANIZATION

In the 1930s, BKL had been divided into six separate collections, each with its own curator. There were separate collections for the mosses, plants native to Long Island, economically useful plants, cultivated plants, Old World plants, and New World Plants other than those from Long Island. In the 1970s a curator began to integrate the collections, but only completed the ferns. Most of the specimens had not been studied for many years so that the identifications were well out of date.

After much debate, the staff decided to reorganize the collection in a way that would be least intimidating to non-taxonomists. Following Mabberley (1987), gymnosperm and flowering plant families are arranged alphabetically within their respective group, with genera alphabetically within families, and species alphabetically within genera.

This reorganization was carried out in conjunction with the move of the herbarium to the new facility. Movers wrapped the specimens in each shelf of every case into a bundle, as you would for a loan, and wrote on each bundle the proper case number and letter. Each full case of specimens was paired with a clean empty case to relieve overcrowding and allow some space for collection growth. After freezing, the coded bundles from a single case were unwrapped and the specimens put into coded shelves in the two new cases according to the new alphabetical arrangement. This crude rearrangement was refined as the collection was treated and unpacked as described earlier.

At one time, the curators could not reliably locate a specimen in the herbarium. Now, even persons with no taxonomic training can find specimens without difficulty. As a result, the use of the collection has increased and it has become a more useful reference for gardeners, ecologists, Garden members, and taxonomists.

CONCLUSION

In our work we have found, first, that good cases are the first line of defense against most of the problems that affect collections. The cases need to be of heavy-gauge steel, well-sealed and light-colored. White powdercoat finishes are preferable to paint because they help diffuse the light and make it easier to spot evidence of insect infestations. Second, it is possible to eliminate insect pests by controlling the herbarium environment. We found that initially freezing the specimens to kill pests in the previously infested collections, in combination with maintaining the herbarium at about 62°F (18°C) for an extended time period, eliminated most insect pests in the collection in a little more than two years. It is hoped that lowering pollution levels and controlling humidity will also facilitate the elimination of additional sources of more long-term damage, especially atmospheric pollutants. Finally, we recommend that new facilities should be built so that they can be easily adapted and upgraded. In the few years since the building renovation was designed, our understanding of herbarium curation has expanded and we have had to adapt our facility and our procedures to keep up to date.

We are pleased with our new herbarium. We have been able to take an excellent but poorly maintained collection and rehouse it so that it is safer, more accessible, and more useful. We realize that we will be constantly changing and improving our herbarium, but that is part of our job as curators, to provide the best environment for the specimens in our care.

ACKNOWLEDGMENTS

I thank William P. Lull of Garrison/Lull Inc., Princeton Junction, New Jersey, and Catherine Hawks of Falls Church, Virginia, for their help in setting and evaluating the parameters for our conservation environment. I thank Linda Marschner, Herbarium Collection Manager at Brooklyn Botanic Garden, for her work, especially in regard to the maintenance and evaluation of the new herbarium. I thank Dr. Steven Clemants and Dr. Stephen K.-M. Tim, my colleagues at Brooklyn, for their ideas, help, critical evaluations, and support throughout the design, construction, and evaluation of the new herbarium. I also thank Deb Metsger and Sheila Byers of the Royal Ontario Museum and Barbara Moore of the Peabody Museum of Natural History, Yale University for their splendid work organizing the workshop, "Managing the Modern Herbarium", held at the University of Toronto, June 5 and 6, 1995.

LITERATURE CITED

BAER, N. S. AND P. N. BANKS. 1985. Indoor air pollution: effects on cultural and historic material. International Journal of Museum Management and Curatorship 4:9–20.

BRIDSON, D. AND L. FORMAN. 1992. *The Herbarium Handbook*. Revised edition. Royal Botanic Gardens, Kew.

BRIGGS, D., P. D. SELL, M. BLOCK, AND R. D. I'ONS. 1983. Mercury vapour: a health hazard in herbaria. New Phytol. 94:453–457.

CLARK, S. H. 1986. Preservation of herbarium specimens: an archive curator's approach. Taxon 35:675–682.

FLORIAN, M.-L. 1986. The freezing process — effects on insects and artifact material. Leather Conservation News 3:1-14.

FLORIAN, M.-L. 1990. The effects of freezing and freeze-drying on natural history specimens. Collection Forum 6(2):45–52.

GILBERG, M. AND A. BROKERHOF. 1991. The control of insect pests in museum collections: the effects of low temperature on *Stegobium paniceum*

(Linnaeus), the drugstore beetle. Jour. Amer. Inst. Conservation 30:197–201.

GILBERG, M. AND A. ROACH. 1991. The use of a commercial pheromone trap for monitoring *Lasioderma serricorne* (F.) infestations in museum collections. Studies in Conservation 36:243–247.

HAWKS, C. A. 1993. 1993 IMS Conservation Assessment. Herbarium, Brooklyn Botanic Garden. Report to the Garden.

LAFONTAINE, R. H. AND P. A. WOOD. 1982. Fluorescent Lamps. Revised edition. Canadian Conservation Institute Technical Bulletin 7:1–10.

LULL, W. P. 1991. Conservation Environment Consultation and Suggested Program for Conservation Environment Renovation. Report to Brooklyn Botanic Garden.

LULL, W. P. 1992. Selecting fluorescent lamps for UV output. Abbey Newsletter 16(4):54,55.

LULL, W.P. AND MOORE, B. 1999. Herbarium building design and environmental systems. Pp. 105–118 in Metsger, D. M. and S. C. Byers, editors. *Managing the Modern Herbarium.* Society for the Preservation of Natural History Collections, Washington, DC, xxii+384 pp.

MABBERLEY, D. J. 1987. *The Plant Book.* Cambridge University Press, Cambridge.

PELTZ, P. AND M. ROSSOL. 1983. Data Sheet: Safe Pest Control Procedures for Museum Collections. Center for Occupational Hazards, New York.

STRANG, T. J. K. 1992a. A review of published temperatures for the control of pest insects in museums. Collection Forum 8(2):41–67.

STRANG, T. J. K. 1992b. Museum Pest Management. Canadian Conservation Institute, Ottawa.

WILLIAMS, R. S. 1989. The Beilstein Test. A Simple Test of Screen Organic and Polymeric Materials for the Presence of Chlorine. Canadian Conservation Institute, Ottawa.

WOODS, J., editor. 1983. *Air Quality Criteria for Storage of Paper-Based Archival Records.* National Bureau of Standards, NBSIR 83-2795.

CHAPTER 9

Moving Herbaria:
A Case Study at Berkeley

BARBARA ERTTER

Abstract.—Several techniques were used to transport 1.8 million specimens during the several moves involved in the ten-year renovation of the herbaria at the University of California at Berkeley. During the interim moves, specimens that were to be transferred into newly purchased cases were sequentially transported in wooden half-height cases fitted with wheels. During the final move when only existing cases were available, an initial pool of empty cases was created by temporarily transferring the contents to other cases or boxes. Once the empty cases were installed on compactor carriages, the contents of other cases were sequentially transferred, thereby replenishing the pool of empty cases so that the process could continue. Specimens were transferred between buildings inside the original cases, after being tightly packed in the lowermost shelves.

INTRODUCTION

One challenge that inevitably arises when new herbarium facilities become available is how to move specimens. This challenge was met in different ways during the several moves involved with the ten-year renovation of the University (UC) and Jepson (JEPS) Herbaria at the University of California at Berkeley. Historically, UC and JEPS were maintained as two separate entities, each with its own taxonomic sequence. As a precursor to renovation, most of the eastern hemisphere spermatophytes (the Old World specimens) were pulled out of the UC collection to alleviate overcrowding and placed in their own separate sequence. The remaining UC-JEPS specimens were then amalgamated into a single sequence and the eastern hemisphere material was also reintegrated during the final move. A summary of how these moves were accomplished is presented as a case study for reference by anyone else faced with the task of relocating into new facilities.

MOVING THE COLLECTIONS TO
TEMPORARY QUARTERS

The UC-JEPS renovation took place in several phases, including moves into interim facilities. The first phase of renovation involved the transfer of most eastern hemisphere spermatophytes to an off-site annex on the adjacent Clark Kerr campus. The specimens were moved into 250 newly purchased herbarium cases, thereby alleviating severe overcrowding. To facilitate the move, existing wooden half-height cases were fitted with wheels, clearly numbered, and used as transfer cases. Folders of specimens to be moved were loaded sequentially into the transfer cases, which were then trucked to the Clark Kerr campus where the contents were unloaded into the new cases. The transfer cases were color coded into three groups. At any given time one group was being loaded, one unloaded, and one was in transit. Wear and tear reduced the number of functional transfer cases during the course of the move, but this system otherwise worked quite smoothly.

The second phase of renovation involved the transfer of the remaining bulk of the collection to interim quarters at the Marchant (SCM) building several miles from campus. This move was accomplished in a similar manner, using the same wheeled transfer cases. This was possible because more new cases had been purchased, not for expansion but because the casework in the old herbarium consisted largely of substandard, built-in, multiple-case units that could not be dismantled for reuse without seriously compromising their pest-worthiness.

MOVING THE COLLECTION INTO THE RENOVATED SPACE

The final move, from interim quarters to renovated quarters in the Valley Life Sciences Building (VLSB), was considerably more complicated. Only existing cases were available, and the eastern hemisphere material was to be reintegrated into the main collection as part of the transfer process. The majority of cases would also be installed on compactorized mobile units.

Planning

Planning for the move began several months in advance. The eastern hemisphere genera at the Clark Kerr annex were inventoried, and corresponding droptags were inserted in the spots where they would be integrated into the main UC-JEPS sequence. Loan droptags were verified in both sequences, and case space for material on loan was allocated as required by inserting space-reserving droptags.

The approximately 1,200 herbarium cases were also inventoried as to color, handle type, and handle position. The recently purchased cases all had recessed handles, but the existing pool of cases included many with protruding handles. In the planned compactorized arrangement, dictated by existing columns, only certain banks were wide enough to accommodate protruding handles. We also wanted to have the various colors (beige, white, green, gray) of cases grouped in different parts of the collections area. A great deal of planning therefore went into determining a transfer sequence that would result in the correct cases being available for installation as needed. For example, the phycology collection was transferred as a break in the delivery of the angiosperm sequence, because those cases were needed next.

Transporting Specimens

The actual transfer of herbarium cases and specimens from SCM into VLSB used essentially the same technique that had proved successful at the University of Texas at Austin (TEX-LL), such that most specimens were transported inside the herbarium cases. To minimize jostling during transport, the specimens were shifted into a tightly condensed arrangement toward the bottom of the case, thereby also lowering the center of gravity of the loaded case. Crumpled newspaper was used to fill the remaining space in any partly filled shelves. Because the specimens were packed almost as well as they would be for shipping, the cases could even be tipped and transported on their sides without noticeable damage.

This compaction technique was not used for boxed specimens such as fungi, which would have required individual padding, and other specimens deemed too fragile for such treatment. Rather, these specimens were loaded either into existing half-height herbarium cases that could be easily moved in an upright position, or into the same wheeled wooden transfer cases that had been used for the previous moves. Fluid-preserved material received special packing and handling, dictated by University requirements for hazardous materials.

"Leap-Frog" Use of Existing Cases During Installation

Herbarium cases were empty while being placed on the compactor carriages and seismically anchored, thus reducing the risk of damage to specimens, cases and installers. To provide an initial pool of empty cases for installation, the contents of nearly 100 cases were either combined with the contents of other cases or, if unaccessioned (e.g., processing backlog), packed into boxes. The emptied cases were then transferred to a staging area in VLSB, from which the installers could select arrays with appropriate colors and handles.

Once the initial pool of emptied cases was installed on the compactor carriages, full cases were transferred to the staging area, as many as could fit at a time. Beginning with the gymnosperms, the previously numbered cases were taken, in order, from each of the two sequences (SMC and the Clark Kerr annex). The wheeled wooden transfer cases were then used to transport specimens from both sequences to the installed cases for insertion. The nearly emptied cases were in turn installed on the compactor carriages, making room in the staging area for the next delivery of full cases.

This "leap frog" loading of specimens and cases continued through the spermatophyte, pteridological, and phycological sequences.

As a keystone to the procedure, the specimens transferred into the last of the installed cases were the mycological collections and miscellaneous other holdings that had been temporarily placed in half-height cases, wheeled wooden transfer cases, and boxes. Before being loaded, these specimens were frozen to destroy any potential pests.

SUMMARY

Planning and preparation for the final move took approximately three months. In spite of the added complexity, taking advantage of this opportunity to reintegrate the eastern hemisphere specimens as part of the move, "zipping up" the two sequences, simplified what would otherwise have been a major logistic challenge. The actual move of 1.8 million herbarium specimens was accomplished in approximately two months, interrupted by winter break. Existing curatorial staff (pared by downsizing) and students were supplemented by two grant-funded assistants, hired for a six-month period.

As a final note derived from experience, a system of checkpoints is highly recommended in any major rearrangement of specimens. Deviations from the planned specimen order or case arrangement can thereby be discovered and corrected or compensated for before the entire rearrangement has been completed.

ACKNOWLEDGMENTS

At the risk of slighting any of the many persons involved with the multiple moves throughout the ten-year period, whose contributions are sincerely appreciated, I nevertheless want to give special thanks to the state- and grant-funded support staff whose uncomplaining efforts brought the years of planning to fruition: Deborah Averett, Mary Borland, Laurie Johnson, Brian Knave, Steve Lessica, Richard Moe, Tony Morosco, Elizabeth Neese, Michelle Seidl, Fosiee Tahbaz, Tom Tang, and Margriet Wetherwax. A special mention goes to Walter Appleby, whose own promising career as Collections Manager at Bishop Museum Herbarium was cut short by cancer. Funds for mechanical-assist compactor carriages, cases for the Clark Kerr annex, and technical assistance for the various moves of the UC-JEPS Herbaria were provided by National Science Foundation Grants BSR-84-17804, BSR-89-07439, and DEB-91-319871.

SECTION III

PREVENTIVE CONSERVATION APPROACHES

TO HERBARIUM MATERIALS AND METHODS

CHAPTER 10

Paper Conservation and the Herbarium

GREGORY J. HILL

Abstract.—Herbaria have evolved standards and uniform methods for mounting, housing, and shipping specimens. The procedures and materials used reflect the historical role of herbaria as voucher collections for taxonomic and floristic research. Herbarium specimens are traditionally mounted on paper, which is compatible with plant specimens due to their similar chemical makeup. The paper conservation community has developed standards and procedures for treatment, care, and storage of paper-based objects. At times, some of the procedures developed for the herbarium may seem at odds with the conservation community's "edicts" and conversely, these "edicts" may sometimes be at odds with the requirements of the collections. A basic introduction to paper chemistry and paper conservation is provided, addressing a few of the issues, both unique and general, that surround the use and conservation of paper in the herbarium.

INTRODUCTION

From the earliest days of collecting botanical specimens, paper has been the preferred mounting substrate. Until the 18th century, specimens were typically mounted in bound volumes. This practice was generally abandoned in favour of flat sheets for ease of examination and arrangement, and for greater security. Paper is essentially a very compatible storage and mount material for plant specimens for a number of reasons: both are cellulose based; both are vulnerable to many of the same agents of deterioration; and both react in a similar manner to fluctuations in relative humidity (RH) and temperature. What differentiates the two is the presence of additives that affect the chemical stability of paper. The intent of this paper is to provide an overview of the composition of paper and related agents of cellulosic deterioration; to discuss specific problems and possible solutions for herbaria from the perspective of an archival conservator; and to ultimately assist the botanist in making informed decisions when selecting mounting and storage materials and methods.

COMPOSITION OF MODERN PAPERS

Rags are brought unto my mill
Where much water turns the wheel
They are cut and torn and shredded,
To the pulp is water added;

Then the sheets 'twixt felts must lie
While I wring them in my press.
Lastly, hang them up to dry
Snow-white in glossy loveliness.

     This poem, written in 1576 by Hans Sachs (Hunter, 1930), very nicely encapsulates early paper-making technology. It is still possible to find papers from this period that remain "..snow-white in glossy loveliness," but unlikely for papers made in the latter half of the 19th century and the 20th century.

     Composition and processing are key to chemical stability of papers. Historically, paper was fabricated from cotton or linen fibres which are typically pure, chemically stable, and strong. If made from similar fibres, contemporary papers share these characteristics. However, the primary fibre used in the commercial paper industry for the past 125 years has been extracted from hardwood and softwood trees. A rigorous purifying process is required to make these fibres chemically stable, but, due to cost, this is not commonly done. The explosion of printing technologies in the early 20th century fueled the demand for paper, a demand readily met with these inferior quality wood pulp papers. We now find ourselves in serious trouble in our fight to preserve our written heritage due to the rapid deterioration of these papers. Further, many herbaria may be in the position of having to replace many of the papers from the 19th and early 20th century that were used in the mounting and storage of specimens.

The production of modern commercial papers involves grinding then stewing the wood to separate the fibres. Prior to paper formation, extractives (including minerals, resins, and gums) are removed leaving three major constituents: cellulose, hemicellulose, and lignin.

Cellulose occurs in nature principally as a hollow, elongated fibre in the walls of plants. The building block of the cellulose molecule is a glucose unit which forms into chains of varying lengths according to the type of fibre. Generally speaking, the longer the chain length, the more resistant the cellulose molecule is to degradation (Whitney, 1979). Cotton cellulose, for example, has a very long chain length and is, therefore, a stable fibre from which good quality papers are made. There are several different types of cellulose. Alpha-cellulose is a type of cellulose that has proven to be more resistant to deterioration under a wide range of test conditions and is therefore preferred for stable papers (Casey, 1960-61).

Hemicellulose is another type which is similar in composition to cellulose but has a much shorter chain length and is therefore less resistant to degradation. However, hemicellulose bonds more readily than cellulose so its inclusion may be desirable in some papers.

Lignin is defined as the glue that sticks the cellulose fibres together. It is generally considered to have no set chemical composition with an amorphous structure that varies from plant to plant. Lignin is an undesirable component in paper making, as it is a source of acidity that will adversely affect the chemical stability of the cellulose fibres.

The proportion of these three major components of paper varies greatly according to the plant source. Cotton has a very low percentage of lignin and hemicellulose in contrast to either softwood or hardwood pulp (Table 1). Flax is in between cotton and wood pulp. Typical hardwood pulp is only slightly more stable than softwood pulp. The acceptable level of lignin is widely debated between preservation specialists and scientists on one side and the paper industry on the other. The removal of lignin from wood pulps is not an easy task and adds considerably to the expense. Lignin is presently being investigated in a Canadian cooperative research project, developed and funded by a number of federal and provincial government agencies and by members of the pulp and paper industry. The project is being carried out by the Canadian Conservation Institute (CCI) and the Pulp and Paper Research Institute of Canada (PAPRICAN). This research is intended to determine whether lignin negatively affects paper permanence and, if so, whether, as lignin is known to react with environmental pollutants, this effect can be counteracted by an alkaline reserve. Currently, however, papers with high lignin content are considered unsuitable for archival purposes. The American National Standards Institute (ANSI) revised their standard No. Z39.48 on paper permanence in 1992; it calls for a lignin content of no more than 1% (ANSI, 1992).

TABLE 1. Approximate composition of the three major components of cellulose (extractive-free basis).

	Cotton linters	Raw flax	Typical softwood	Typical hardwood
Cellulose	96%	85%	50%	50%
Hemicellulose	3%	10%	20%	30%
Lignin	1%	5%	30%	20%

after DePew, 1991

If the presence of lignin is suspected in a mounting or storage paper the Phloroglucinol test can act as a simple lignin indicator (Barrow, 1969). A drop of the solution is placed on a disposable sample of the paper. If the spot turns purple or magenta, the fibre contains lignin. If the spot remains colourless or becomes slightly yellow, no lignin is present.

PULPING PROCESSES

Besides the inherently unstable nature of various raw materials used in paper making, deleterious acidic components may be introduced during the various stages of pulp preparation. There are two basic pulping processes in general use: mechanical pulping and chemical pulping. Mechanical pulping separates the fibres by physical abrasion, resulting in weak, short-fibred papers with little or none of the lignin and other impurities removed. Newsprint is a prime example of this process. Chemical pulping uses either strong acids or strong alkalis to break down and purify the cellulose and produces a chemically unstable paper with varying degrees of impurities (including lignin) and varying strength.

SIZES

Sizes are added to pulps to strengthen the paper and to make the surface suitable for accepting ink. Without them, paper would act like a blotter allowing the inks to feather and sink deep into the paper fibre. Historically, sizing agents have included: starch, gelatin, various rosins, and a variety of alum/rosin and alum/gelatin combinations (Bruckle, 1993). Their impact on the cellulose can be severe; alum in particular is quite acidic. Today there are many sizing compounds and systems which make it possible to size paper to any acidity or alkalinity and any degree of hardness. Aquapel, for example, is a new synthetic sizing agent that is alkaline, stable, and very effective (Whitney, 1979). Most acidic commercial papers are still sized with alum/rosin combinations which give the paper a pH of 4.5 to 5.5. The Aluminon test for aluminum ions can be used to detect the presence of alum (Barrow, 1969). Again this is a spot test applied to a test sample of the paper.

FILLERS/COATINGS

Over the years, a variety of coatings and/or fillers have been applied to printing papers in an attempt to provide smoother, more pristine surfaces for inks. These papers are commonly found in magazines and books with photographic reproductions. Standard fillers such as calcium carbonate, kaolin clay, and titanium dioxide are usually bound to the paper using gelatin or various synthetic binders (Whitney, 1979). Most of these coated stock materials have ligneous paper cores that are acidic. All are very problematic when wet or in a very humid environment as they tend to stick solidly together.

OTHER PAPER ADDITIVES

Other materials which are routinely found in papers, and which can affect the long term stability of the cellulose, include optical brighteners, modified celluloses, starch, and wet-strength resins.

Bleaches

Chlorine was isolated and identified in 1774. This soon lead to the development of chlorine bleaches still widely used today (DePew, 1991). Although a very effective bleach, chlorine is difficult to thoroughly remove from paper fibres. Consequently chlorine residues combine with water to form hydrochloric acid, which hydrolyses the cellulose.

Dyes/Colorants

Almost all papers, including those that are already white, are dyed to ensure uniform colour (DePew, 1991). Colorants vary: some are dyes, some are pigments, and some are combinations of the two. Almost all are subject to photochemical degradation, primarily fading.

EXTERNAL AGENTS OF DETERIORATION: ENVIRONMENTAL FACTORS

The discussion of the components of paper has so far concentrated on the problems inherent in the paper. This section will examine the way in which these potential and real problems manifest themselves and are exacerbated by the environment.

The principal agents of paper deterioration are physical,

chemical, biological, biochemical, and photochemical. Some agents damage objects or specimens directly and others react with their inherent problems.

Physical

Handling is one of the greatest single causes of deterioration of paper. Safe handling is critical to avoid physical damage to specimens mounted on paper. Shipping of specimens or objects requires a packaging system that will neither apply too much pressure to them, nor allow any movement or flexing of them. When specimens are removed from storage shelves, the rigidity of the mounting paper should not be overestimated. Specimens must be supported from below by a hand or, preferably, a full rigid support. As paper is weakened by creasing and flexing, mounted plant specimens are at increased risk, especially if the mount is made of poor quality, acidic paper.

Chemical

The chemical stability of paper is the key to its long-term survival. Sources of chemical instability (essentially acidity) can be internal or external. External sources of acidity include air pollution, poor quality storage materials, and handling. Acid hydrolyses the cellulose molecule causing the paper to lose its strength; the polymer chains gradually break down and the paper becomes weak, brittle, and discoloured. In the presence of even low levels of humidity, acids can migrate from one object to another, making poor quality storage materials a potential source of acid. In humid conditions, sulphur-based airborne pollutants form sulphuric acid which, although dilute, gradually deteriorates most organic and inorganic materials. Other damaging pollutants include nitrogen dioxide (which forms nitric acid) and various oxidizing agents such as peroxides and ozone. Acid transferred from skin during handling can also cause hydrolysis of the cellulose fibre.

Biological

Plant materials, by virtue of being organic, are subject to attack by insects, molds, and rodents. Mold spores are ever present in the air, remaining dormant under dry conditions. In high humidity (above 70%) and particularly with elevated temperatures, they rapidly grow, attacking sizing agents, paper fibres, and surface media. The effect is weakening and staining of the paper and, if unchecked, total destruction. Insects also prefer elevated temperatures and humidity and can cause total destruction of collection material. Organic materials are also potential food supplies for rodents, particularly during winter months.

Biochemical

Biochemical degradation of paper generally takes the form of disfiguring yellow or brown spots known as foxing. It has been attributed to trace metal particles embedded in the paper fibre and/or mold, often a combination of the two. Again, high humidity plays a major role in its formation.

Photochemical

All organic materials are subject to photochemical degradation. Light or radiant energy both in the visible spectrum and in the ultraviolet range can affect very sensitive materials, speeding up oxidation reactions and chemical breakdown. It is the non-visible, ultraviolet range (wavelengths between 100-400 nanometres) that tends to be more damaging. The primary sources of UV light are sunlight and fluorescent tubes. A newspaper left in sunlight for a day, for example, exhibits a yellowing or darkening of the paper as the lignin reacts with other components of the paper. In herbaria, however, photochemical degradation may result in fading or bleaching not only of the paper, but also of the pigments, dyes, or any plant specimens that have been mounted on it. Although not a critical problem for all herbaria it may be for some. Even so-called archival quality materials are subject to fading and colour change in the presence of strong light.

INTERACTION OF ENVIRONMENTAL FACTORS WITH AGENTS OF DETERIORATION

The environment plays a pivotal role in the deterioration of all papers by any and all of the agents just discussed. Oxidation reactions, regardless of the cause, speed up in the presence of high humidity, particularly if accompanied by high temperatures. High temperatures with low levels of humidity can desiccate cellulose fibres and cause embrittlement. Warm, humid conditions (above 70% RH) can promote mold growth.

The acceptable temperature range for storage of paper is 18 to 20°C with RH ranging from 35% to 55% (Canadian Council of Archives, 1990). They should be as stable as possible within those ranges with fluctuations not exceeding 2°C or 3% RH.

ISSUES RELATED TO THE HERBARIUM

The following issues relate specifically to the herbarium and its use of paper. Clearly, procedures and working methods within the herbarium have been developed over a long period of time to suit the requirements of the collections and the people using them. The intent here is not to suggest changes to procedures, but rather to acknowledge potential problem areas from a conservation perspective, and to suggest possible solutions.

Mounting Papers

Paper has been an integral part of the herbarium from virtually the beginning. Using high quality, chemically stable papers provides a considerable degree of protection for the collections. Most plant samples are acidic by nature. Using an alkaline buffered mounting paper will therefore promote greater long-term stability of the herbarium specimen, as the buffer in the paper will neutralize the acids migrating from the plants. However, as certain highly acidic plant species may be adversely affected by an alkaline environment, careful consideration must be given to the use of buffered papers. Until recently, a common misconception about buffering, or an alkali reserve, has been that, unlike acids, it does not migrate. Recent research has shown that alkali reserve does, in fact, migrate and that this migration is humidity dependent (CCI, pers. comm., 1995). Therefore, controlling the RH in the storage environment will greatly reduce the rate of migration. Continued discussion on this issue is required between conservators and herbarium staff. Ultimately, substantial research is needed to further examine the effect of alkalinity on plant specimens.

ANSI standard Z39.48-1992 (Standard for Permanence of Paper for Publications and Documents in Libraries and Archives) stipulates that permanent paper should be 100% alpha-cellulose, acid- and lignin-free (maximum 1%), should have neutral or alkaline sizes and should incorporate a minimum deposit of 2% calcium carbonate alkaline buffer. A cold extraction pH test of a commercially available herbarium mounting sheet indicated a pH of 9.3. This is slightly higher than that normally suggested for enclosures for general paper based collections (Canadian Council of Archives, 1990), and would be considered unacceptable for alkali-sensitive items such as some plant and some photographic materials. To keep things in perspective, however, the impact of the pH can be minimized by controlling the RH. A sheet such as this may still be preferable to a poor quality, highly acidic paper mount.

Packaging Specimens for Travel

Many herbarium specimens travel on loan. This can be a risky undertaking as most are transported by mail or courier. The almost universal materials and procedures for packaging specimens is to: wrap the individual sheets in newsprint; bundle them in batches of up to 25 (to a maximum thickness of approximately 2 inches [5 cm]); place them into double folders (two folders end to end and taped); sandwich each folder package between corrugated cardboard sheets for rigid support; and place the bundles, one on top of the other, in a corrugated cardboard box (some herbaria use double walled boxes). The stacks within the box are then secured with a filler, such as crushed paper, bubble pack, or occasionally styrofoam chips. The box is then sealed with pressure-sensitive packing tape. Finally, the boxes are wrapped with paper and sealed a second time with packing tape along the seams.

Physical damage from mishandling is a major threat throughout the loan process, as is chemical instability due to adverse environmental conditions imposed by the packing materials and procedures described. Bagging specimens in airtight, stable plastic bags, such as food freezer bags, greatly reduces the environmental risk factor involved in shipping specimens for loans. Cost is understandably a major factor, so using reusable packing materials cut to standard sizes is advisable if not already being done. Ordering supplies in bulk between a number of institutions can also substantially reduce costs. Ink transfer from printed newspaper is a common problem. Using unprinted newsprint is preferable and relatively inexpensive.

Photographic Materials

Photographic materials attached to specimen mounting sheets cause a variety of problems. Photographic print materials are traditionally made from very high quality chemically-stable papers. Photographic emulsions, on the other hand (particularly those of historic and colour materials), can be rather fragile. Removing these materials and storing them separately, where possible, is therefore recommended. When there are no negatives, replacing original prints with facsimiles is a good alternative if resources permit. One of the potential problems facing photographic emulsions attached to specimen sheets is the alkaline buffering of the mounting sheets. Photographic materials are acidic by nature and may be adversely affected by alkaline storage environments. The degree to which they are sensitive to alkaline materials has not yet been determined though most specialists in the field agree that neutral, acid-free storage materials are preferable to alkaline ones (Hendriks et al., 1991). The fact that most commercially available mounting sheets are highly buffered, is particularly problematic for photographic materials under conditions of high humidity. High humidity and high temperatures also greatly affect the stability of photographic dyes, particularly those in chromogenic materials (regular photo-finishing negative and print materials). Although considerably more stable, even Cibachrome dyes are subject to dye loss under extreme environmental conditions. The method by which photographs are attached to specimen mounting sheets must also be considered. Commercially available polyester photo-corners are the simplest solution to the problem of mounting as adhesives or adhesive tapes must not be used directly on the photographic print.

Folders for Type Specimens

Type specimens are valuable and rare and demand special treatment. Rigid folders that provide physical protection from all sides should be fabricated from acid-free, chemically stable materials such as museum mounting board (Shchepanek, this volume). Depending on budget and the requirements of the specimen, both buffered and non-buffered boards should be used, buffered for general plant materials without alkali-sensitivity and non-buffered for alkali-sensitive plant materials.

Light, Temperature, and Relative Humidity

Monitor the storage environment! Equipment such as hygrothermographs, psychrometers, dataloggers, LUX meters, etc. are readily available. Should budgets preclude the acquisition of this equipment, various agencies such as other museums, galleries, and conservation research centers like CCI may be able to supply the required expertise and equipment, on a regular if not frequent basis. In the event of high humidity, dehumidify by whatever means available to avoid cockling of the mounting papers and to eliminate the potential for mold and insect damage. Lowering the temperature and RH will result in greater chemical stability for most organic materials, thereby extending the life, in this case, of the paper. A general rule of thumb for chemically unstable materials such as typical wood pulp papers is that, for every 5°C drop in temperature, the life span of an object or specimen is doubled (Michalski, 1989).

Most herbarium specimens are stored in the dark, thereby minimizing light damage to them. This is not the case, however, for specimens that go on display or are being examined for long periods of time. Light damage is cumulative and non-reversible and can result in colour loss and oxidation. However, besides fading, no actual research has been conducted to determine other effects of light on plant fibre.

Storage Cabinets or Cases

Cabinets or cases should be constructed of stable materials that will not release potentially damaging gases. Due to cost, untreated wood shelves may be all that is available. Coating them with high quality acrylic latex paint will reduce, though not stop emissions of damaging vapours from the wood. Theoretically, in this scenario, ensuring that the air exchange rate within the cabinet is greater than the rate of gas release will minimize the problem. Providing some ventilation or air exchange and placing a barrier of acid-free buffered paper or Mylar (polyester) between the specimen and the shelf would further reduce potential damage (Slavin and Hanlan, 1992).

Storage Environments

Clean, controlled environments are the best defence against insect and mold attack. The use of most insecticides and biocides should be discouraged due to the health risk and their effect on both paper and plants. In the past some herbarium specimens were treated with mercuric chloride to counter insect infestation. The mounting sheets were often left with a greyish tone. These specimens are a health and safety hazard due to the presence of residual mercury and should be removed from collections. Gloves should be worn when handling them.

Newsprint Storage

In spite of the availability, price, and convenience, long term storage of unmounted specimens in newspaper is not recommended due to its highly acidic nature. Printed newspaper is likely to transfer inks to the specimens and therefore unprinted paper should be used.

Fasteners

The use of paper clips and fasteners, plastic or metal, should be avoided. Not only can they cause physical damage to the paper, but in high humidity, metal will easily oxidize, causing staining and degradation of the cellulose.

Post It Notes®

Although the convenience of *Post It Notes®* is well recognized, they leave an adhesive residue behind which is a potential source of acidity and can attract dust and dirt (Grace, 1990). Some paper colorants used in *Post It Notes®* are water soluble and can transfer dyes.

Paper Repairs

For repairs to mounts or labels, use stable adhesives such as wheat or rice starch, or methyl cellulose together with high quality Japanese tissues which are chemically stable, long fibred, and strong. When mounting sheets have deteriorated, plant specimens can be removed and re-mounted onto new sheets. However, fragile specimens may be at great risk, making it preferable to repair the original mount. It may be necessary to have professional paper conservators do this work. They should always be consulted for historic collection material when the mount or album page has potential value as an artifact. Do not succumb to the temptation to use commercially available pressure-sensitive tapes unless you are assured of their chemical stability. Never use staples as they damage the paper and rust in high humidity.

Photocopies

As it is occasionally necessary to replace specimen mounts, it is often desirable to photocopy inscriptions. Photocopying should be carefully done on electrostatic or Xerographic plain paper copy machines that employ carbon black toner. Photocopy paper should conform to the requirements of permanent paper (Australian Archives, 1993).

Dry Cleaning

Surface dirt can be hygroscopic and a potential source of acidity. Heavy layers can be removed with the very gentle use of white vinyl erasers and powdered eraser products. However, to avoid serious physical damage, extreme caution should be exercised with fragile papers. In many cases this job is better left to the experienced conservator. Following dry cleaning, all residue should be removed due to the rather acidic nature of all eraser products (Pearlstein et al., 1982).

CONCLUSION

The purpose of this paper has been to provide an introduction to the study of paper, its composition, chemistry, and conservation, and to connect this with its use in the herbarium. This is a very brief overview. It should be understood that strict standards do exist, not only for the composition of permanent papers, but also for its handling and storage, and for how it should be treated by a conservator. The similarities in composition between paper and plant specimens might foster a false sense of security regarding the conservation of these two materials. As exemplified by the issue of alkaline sensitivity, one cannot always assume that all the rules can be applied from one to the other. The paper conservator can certainly advise on the conservation of paper in the collections and basic storage and housing considerations. Beyond that, careful dialogue is essential between herbarium staff and conservators to address the unique and very specific concerns of the herbarium.

ACKNOWLEDGMENTS

The author would like to acknowledge and thank the collections management and conservation staff of the Herbarium of the Canadian Museum of Nature for their assistance in the preparation of the initial presentation and of this paper.

LITERATURE CITED

[ANSI] AMERICAN NATIONAL STANDARDS INSTITUTE. 1992. Permanence of Paper for Publications and Documents in Libraries and Archives: ANSI Standard Z39.48.

AUSTRALIAN ARCHIVES. 1993. Photocopying and Laser Printing Processes: Their Stability and Permanence. Australian Archives Leaflet: September 1, 1993.

BARROW, J. W., RESEARCH LABORATORY, INC. 1969. *Permanence/Durability of the Book. VI: Spot Testing for Unstable Modern Books and Record Paper.* Richmond, Virginia, pp. 10-13.

BRUCKLE, I. 1993. The role of alum in historical papermaking. Abbey Newsletter 17(4):53-57.

CANADIAN COUNCIL OF ARCHIVES. 1990. *Basic Conservation of Archival Material: A Guide.* CCA, Ottawa, p. 57.

CASEY, J. P. 1960-61. *Pulp and Paper: Chemistry and Chemical Technology, Vol 1.* Second edition. Interscience Publishers, New York, p. 358.

DePEW, J. N. 1991. *A Library, Media, and Archival Preservation Handbook.* ABC CLIO Inc., California, pp. 10-11, 13.

GRACE, J. 1990. Don't Post-it on your archives! The Archivist (National Archives of Canada) May-June, p. 13.

HENDRIKS, K. B., B. THURGOOD, J. IRACI, B. LESSER, AND G. HILL. 1991. *Fundamentals of Photograph Conservation: A Study Guide.* Lugus Publications, Toronto, p. 416.

HUNTER, D. 1930. *Papermaking Through Eighteen Centuries.* William Edwin Rudge, New York, p. 22.

MICHALSKI, S. 1989. *Humidity, Temperature, and Pollution in Libraries and Archives.* Canadian Conservation Institute. Unpublished draft, Feb. 21, 1989, section B.

PEARLSTEIN, E. J., D. CABELLI, A. KING, AND INDICTOR, N. 1982. Effects of eraser treatment on paper. JAIC 22(1):1-12.

SHCHEPANEK, M. J. 1999. A protective hardboard folder for storing valuable herbarium specimens (Abstract). P. 362 in D. A. Metsger and S. C. Byers, editors. *Managing the Modern Herbarium.* Society for the Preservation of Natural History Collections, Washington, DC, xxii+384 pp.

SLAVIN, J. AND J. HANLAN. 1992. An investigation of some environmental factors affecting migration-induced degradation of paper. Restaurator 13(2):78-94.

WHITNEY, R. P. 1979. *Chemistry of Paper.* Conference presentation, International Paper Conference, San Francisco, March 1978. World Print Council, p. 37, 43.

Chapter 11

Adhesive Research at the Canadian Conservation Institute as it Relates to Herbarium Collections

Jane L. Down

Abstract.—The perfect adhesive for herbarium collections is still elusive. Nevertheless, information from adhesive research conducted at the Canadian Conservation Institute (CCI) is available and is summarized here in order to help herbarium staff make informed choices and know limitations when selecting adhesives. To begin, a brief history of adhesive use at the Canadian Museum of Nature Herbaria and a set of conservation criteria for use of adhesives in herbarium collections are outlined. Recent CCI poly(vinyl acetate) research with particular reference to some products that are already in use in herbarium collections is reviewed. A brief overview of the new CCI adhesive project on vinyl acetate/ethylene co-polymer adhesives is given and potential benefits to the herbarium community are highlighted. The suitability of methyl cellulose and a Mylar® strip alternative for mounting plants is discussed. Finally, the subject of labels is examined with particular reference to two pre-gummed labels studied by CCI and to the use of glue sticks to attach labels.

INTRODUCTION

Adhesives are used in herbarium collections to mount plants and to glue labels. On a tour of a herbarium, it is obvious that other concerns such as pest control and the paper used for pressing, mounting, and storing are two major problem areas. The problems associated with the use of adhesives are not always so obvious but important nonetheless. Upon cursory inspection, it can be seen that many adhesives and techniques have been used on herbarium collections in the past, especially for mounting plants, and after decades it would seem that the adhesives are performing their herbarium-related-tasks fairly well. For example, they still hold the plants in place and are not sticky so the mounted specimens can be stacked. But on closer inspection of these collections, some problems with the adhesives come into focus. For example, adhesive strips have broken, adhesives have caused plants to crack, adhesives have stained the mounting paper, adhesives have turned yellow, adhesives can no longer be easily removed from the plant, and what are the adhesives doing to the chemical integrity of the plant especially in cases where the plant has been completely coated with adhesive. In the past, adhesive selection for mounting plants has usually been based on ease-of-use, toxicity, availability, ability of the product to hold, and tack. There are other criteria, *conservation* criteria, that could help herbarium staff think of alternate solutions or, failing that, choose better adhesives in order to avoid some of these problems. The results of *conservation* adhesive research where many of the adhesive problems have been examined could benefit the herbarium community. The Canadian Conservation Institute (CCI) has been conducting conservation adhesive research for the last 20 years. Some of their findings relevant to herbarium collections will be summarized here in order to help herbarium staff make informed choices and know adhesive product limitations. Appendix 1 provides a list of suppliers for many of the materials mentioned.

THE HISTORY OF ADHESIVE USE AT
THE CANADIAN MUSEUM OF NATURE HERBARIA

Let us first start with a brief history of the types of glues that have been used for mounting plants. As a case study, the Canadian Museum of Nature (CMN) Herbaria (CAN, CANA, CANL, CANM) experience is outlined.

In the CMN Herbaria, fish glue and/or paper strips were used in the 1800s to mount plants. Initially, the whole plant was glued from behind onto the mounting paper and sometimes this would cause the leaves to crack after the glue dried. This is not surprising since fish glues tend to be very stiff and brittle. Later, plants were attached using a combination of paper strips for strapping and fish glue. The paper strips did not work well; they tended to tear and required frequent replacement.

Strapping with linen tape was used next from the early 1900s to the early 1950s. Examining the collection today, the linen tape appears to have held well. The adhesive on the linen tape from the early 1900s was found to be a gelatin (Down, 1995a). The adhesive on most of the modern linen tapes sold today is a combination of gelatin and starch (Laver, 1975; Moffatt, 1979 and 1982; Moffatt and Young, 1985) and does not appear to hold as well as the gelatin tape. It is generally thought that starch and gelatin adhesives are fairly stable since they have been around for centuries.

From about 1950 to the early 1970s, an adhesive known as the Archer's formulation was used. This was a mixture of Ethocel™ (ethyl cellulose), Dow Resin 276 V2 (methyl styrene), toluene, and methanol (Archer, 1950). This formulation was toxic and its use was discontinued due to health problems experienced by the preparator. The Archer's formulation contains ethyl cellulose, which we now know tends to degrade and discolour (Feller and Wilt, 1990). For this reason, its use in the formulation would not now be recommended. Methyl cellulose might be a better choice since it performed better than ethyl cellulose (Feller and Wilt, 1990). This performance was based on a water-soluble variety of methyl cellulose; no organic-soluble variety was tested. Since an organic-soluble variety would be needed for the Archer's formulation, further research to determine if organic-soluble methyl cellulose is more stable than ethyl cellulose would be required before a switch to methyl cellulose could take place. Evidence (Feller and Wilt, 1990) suggests that other organic-soluble varieties of cellulose (e.g., ethyl hydroxyethyl cellulose) are not as stable as the water-soluble variety which further emphasizes the need to investigate such a switch carefully. As a further note, the use of methyl cellulose instead of ethyl cellulose in the Archer's formulation could affect other properties (e.g., viscosity, setting time, etc.)

and this would require further investigation before implementation. For those herbaria still using the Archer's formulation, it is absolutely imperative that proper ventilation (e.g., fumehoods) be employed when preparing the specimens and especially during adhesive drying.

Poly(vinyl acetate) (PVAC) emulsions, the so-called white glues, have been used in the CMN Herbaria from the early 1970s and are still in use today. Elmer's® Glue-All was used frequently. PVAC emulsions are definitely less toxic than the Archer's formulation, but Elmer's® Glue-All is not the best choice as it is acidic and yellows (Down et al., 1996). It should also be noted that the formulation of Elmer's® Glue-All has changed over the years (Down et al., 1996) illustrating one of the problems with using commercial products: formulators can change products without notice. Today, the PVAC emulsion Weldbond® is being used in the CMN Herbaria, despite the fact that it is acidic. It does have other redeeming qualities which will be discussed later. The current method of mounting most plants in the CMN Herbaria involves the use of Weldbond® and strapping with gelatin-coated linen tape.

CONSERVATION GUIDELINES FOR USE OF ADHESIVES IN HERBARIUM COLLECTIONS

The criteria for use of adhesives in herbarium collections can be considered under two headings: those that are important from a herbarium point of view and those that are important from a conservation point of view. Considering the herbarium criteria, the adhesive should hold well and immobilize the plant, be non-toxic (i.e., preferably an emulsion product since they are water-based), be easy-to-use, have no dry tack (i.e., not sticky when dried) so sheets do not stick together, and should not puddle but hold its shape (i.e., have high viscosity). While the herbarium criteria are quite straightforward, the conservation requirements and considerations are more fundamental. The first conservation criterion would be to question the use of an adhesive at all. Conservators would first look for suitable alternate solutions that would not put artifacts or, in the herbarium context, plants, in direct contact with adhesives. This would be discouraged since ingredients in some adhesives could accelerate deterioration of the artifact/plant or might interfere with future genetic/chemical testing of the artifact/plant. Alternate solutions for plant mounting might involve strapping, sewing, or using techniques that use much less adhesive. One example of an alternate mounting solution being investigated by CCI which avoids direct contact of the adhesive with the plants is discussed later in this paper (see Mylar® Tape Alternative). If, after a thorough exploration of alternatives, an adhesive treatment is still indicated, then the adhesive should meet certain standards. The adhesive should have a neutral pH and remain neutral with time so that it will not deteriorate the mounting paper or the plant. The adhesive should be fairly flexible and remain flexible so that it does not embrittle and de-adhere. It should not release harmful volatiles such as acids which could deteriorate the plant itself or other specimens close by. It should be non-yellowing. Although yellowing of the adhesive does not seem to be a concern for herbarium collections from an aesthetic point of view, it could indicate deterioration of the adhesive and should be avoided if possible.

In conservation, one major concern is that of being able to reverse a process. We try never to do something to an artifact that we cannot undo or remove later. This is one reason why, wherever possible, we try not to put adhesives in direct contact with artifacts. If this *is* necessary, then we try to use the most stable adhesives to minimize the frequency of future removal and repair, and to use reversible adhesives so that removal is possible. It was interesting to learn that in herbarium collections, one seldom removes the adhesive from the plant or the mounting paper. If the mounting paper disintegrates or the adhesive de-adheres, it is standard practice to re-glue the plant onto another sheet of mounting paper transferring the old adhesive with it in order not to damage the plant. Removal of the old adhesive would not be attempted or considered necessary. In conservation, removal of the adhesive is always considered — not always achieved but always considered.

Finally, if an adhesive is to be used, herbarium staff should consider the amount of adhesive they use. The amount may vary depending on the type of plant and herbarium. For instance, bulky plants may require more adhesive than delicate plants. Some herbaria such as those associated with teaching establishments see their specimens handled more frequently and sometimes more vigorously than those associated with government or private collections. If the mounted specimens are handled often and vigorously, then more adhesive should be used in their preparation than would be the case for specimens that see little handling. The rule should be to use as little adhesive as will hold but enough to allow handling protection. Coating the entire plant would never be recommended.

CCI PVAC AND ACRYLIC ADHESIVE RESEARCH

Recently at CCI, a research project on the stability of PVAC and acrylic adhesives was completed (Down et al., 1996). Many of the conservation criteria mentioned above were examined for the PVAC adhesives. Several of the PVAC adhesives that are already in use in many herbaria were examined in the study. A brief summary of some of the results can be seen in Table 1. This table includes some PVAC adhesives that might be familiar to herbarium staff as well as those that were most successful in the CCI testing program. The latter appear first in the table. They are fairly neutral, release low to medium volatiles, have medium strength and good flexibility, and do not discolour rapidly. Nevertheless, there are problems with these three PVAC emulsions: Jade No. 403 now has a very offensive odour that it did not have in its earlier formulation, R-2258 is no longer available, and Mowilith® DMC2 may dry too tacky for herbarium collections and is now difficult to obtain. The last six PVAC emulsions shown in Table 1 are known to have been used in herbarium collections. All of these emulsions are acidic, some release high quantities of volatile acetic acid, most are fairly strong (probably not a problem for herbaria), some embrittle, and some discolour fairly rapidly. The best of these products is Weldbond® but it has a problem with acidity. CCI has been experimenting with buffering acidic adhesives to make them neutral. For example, Figure 1 shows an emulsion adhesive, Dur-O-Set® E-150, whose pH was adjusted with calcium carbonate. Even a small amount (1%) of calcium carbonate made

Table 1. Overall performance of selected poly(vinyl acetate) emulsion adhesives under different aging conditions.

Adhesive	pH			Volatiles		Strength			Flexibility			Yellowing		
	0	D	L	D	L	0	D	L	0	D	L	0	D	L
Jade No. 403	N	N	N	l	l	M	M	M	F	F	F	G	f	f
Mowilith® DMC2	N	N	N	m	m	M	M	M	F	F	F	G	G	f
R-2258	N	N	A	l	m	M	M	M	F	F	F	G	G	G
Weldbond®	A	A	A	m	m	M	M	M	F	F	F	G	G	G
Gaylord Magic Mend	A	A	A	l	m	M	S	S	F	F	F	G	f	f
Bulldog Grip® 20 Min	A	A	A	m	m	M	S	S	F	F	B	f	f	f
Elmer's® Glue-All	A	A	A	m	m	M	S	S	F	B	B	f	f	f
Bondfast™	A	A	A	m	h	M	S	S	F	B	B	G	f	f
Bulldog Grip® 2311	A	A	A	l	m	S	S	S	B	B	B	P	P	P

For each entry in the table, the worst value measured for each property over the entire aging period appears. The aging period is different for each property and varies from 1 to 5 years, most aging periods are about 4 years (Down et al, 1996).

0 = 0 years;   *D* = dark aging;   *L* = light aging

Shading indicates desirable properties.

pH	A = acidic (pH < 5.5)
	N = neutral (pH = 5.5-8.0)
Volatiles	l = low (< 1µg acetic acid / g adhesive)
	m = medium (1-10 µg/g)
	h = high (> 10 µg/g)
Strength	M = medium (2-15 MPa)
	S = strong (> 15 MPa)
Flexibility	B = brittle (elongation < 20% or modulus > 2000 MPa)
	F = flexible (elongation > 20% or modulus < 2000 MPa)
Yellowing	G = good (A_t < 0.05)
	f = fair (A_t = 0.05 to 0.10)
	P = poor (A_t > 0.10)

quite a difference to the extracted pH of the dry film, increasing it from 4.09 to 7.18. Although adding calcium carbonate to an acidic PVAC adhesive to neutralize it seems to be the solution, to date the aging effect of this addition is unknown. CCI's new adhesive research project, which is described later in this paper, will address this issue.

CCI research (Down et al., 1996) suggests that there is still no perfect PVAC adhesive. At this point, if a PVAC adhesive is to be used, the best compromise for herbarium collections would be Mowilith® DMC2 or Weldbond®. The dry tack of Mowilith® DMC2 could be improved perhaps by the addition of a matting

Figure 1. The effect of adding calcium carbonate to Dur-O-Set® E-150 on the extracted pH of dry films and on the pH of the wet emulsion.

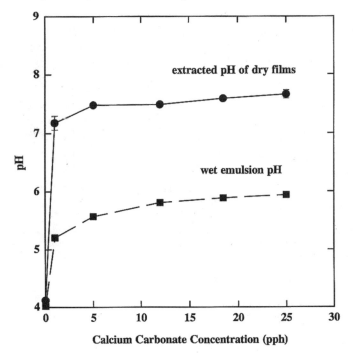

agent or filler. Similarly, buffering Weldbond® with calcium carbonate may improve it. Both these improvement methods require more testing for possible aging effects.

Acrylic adhesives were also studied in the CCI project (Down et al., 1996). In general, the acrylic adhesives displayed better properties (i.e., neutral pH, no harmful volatiles, moderate strength, good flexibility, non-yellowing) than most of the PVAC adhesives and thus should, in theory, present more choices for mounting specimens. To our knowledge, few acrylic adhesives have been used for mounting herbarium specimens. Of those studied by CCI, few would be suitable for the following reasons:

- Many were organic-soluble and thus would pose toxicity problems due to the solvent (i.e., they would require proper ventilation, preferably a fumehood, for use and drying).

- Others were pressure-sensitive adhesives and thus inappropriate, due to their excessive dry tack, for mounting plants by the standard method of drizzling the adhesive over the plant. However, these same pressure-sensitive adhesives could perhaps be used for alternative plant mounting techniques (see Mylar® Tape Alternative) or for labelling.

- The acrylic emulsion products included in the CCI project were very watery and would require thickening before use.

There are other acrylic adhesives that CCI did not study that might be suitable. For instance, Lascaux 498HV is a non-toxic, viscous emulsion that dries relatively tack-free but its pH, yellowing, and aging properties are unknown. Certainly, more exploration of acrylic products is warranted.

CCI VINYL ACETATE / ETHYLENE COPOLYMER RESEARCH PROJECT

In 1994, the newest CCI adhesive research project began (Down, 1995b). The purpose of this new project is to examine the effect of adding known quantities of modifiers to the long-term stability of a neat vinyl acetate/ethylene (VAE) copolymer emulsion adhesive obtained from a manufacturer. In the PVAC research described above, the VAE copolymers, as opposed to the PVAC homopolymers, were identified as potentially good conservation adhesives that warranted a closer look. The Jade No. 403 and R-2258 adhesives were VAE copolymers. In the PVAC project, many *formulated* products were tested. Besides not knowing exactly how much of an additive was present in these formulations or its effect if present or absent, we were never sure if we reordered the formulated adhesive in the future that we would get the same product with the same additives and proportions. Since manufacturers make the adhesive bases that are sold to formulators, obtaining adhesives from manufacturers rather than formulators could eliminate much of this uncertainty. We thus proceeded to locate manufactured VAE emulsions that did not contain any additives so that we could do the addition ourselves under controlled circumstances.

A few manufacturers sell these neat VAE emulsions. They contain no extra additives except those necessary for polymerization and emulsion stabilization (e.g., a protective colloid). These emulsions are well characterized in that the percent ethylene content,

TABLE 2. Evaluation of neat VAE copolymer emulsion adhesives.

Evaluation criteria	Airflex® 320	Airflex® 465	Dur-O-Set® E-150	Elvace® 40705	Elvace® 40709	Elvace® 40724	Jade No. 403
Formaldehyde (%)	<0.05	<0.05	*	*	*	*	trace
Ethylene in copolymer (%)	5-7	17	11	9-12 high	medium	5-7 low	~13
Dry tack	none	tacky	none	slight	very slight	none	very slight
FTIR comments	soap peak	no soap peak	no soap peak	soap peak	soap peak	soap peak	?
Other comments	-	-	-	-	-	too brittle	smells

Shading indicates desirable or tolerable properties.   * = Manufacturer does not mention that formaldehyde is present.
FTIR = Fourier Transform Infrared Spectroscopy.   Dry Tack = The stickiness of a dry film of adhesive.

choice of protective colloid, and amount of hydrolysis are known. Several neat VAE emulsions were examined for this project (Table 2). Dur-O-Set® E-150 was selected for testing for the following reasons:

- conservators thought it was the most similar in film and liquid properties to Jade No. 403,
- it was 'purer' than some of the other neat products in that it contained no formaldehyde or soap,
- it had virtually no smell, and
- it is sold by a Canadian supplier.

pH was not used as a selection criterion because all the products were acidic (pH ~ 4) and would require pH modification before use.

Modifiers will be added to the Dur-O-Set® E-150 in known concentrations. Table 3 lists all the modifiers that have been selected for the project. The modifiers will be added to the Dur-O-Set® E-150 individually, not in combinations. Combinations will be considered later if the results warrant it. Yellowing, pH, flexibility, film strength, gloss, and removability will be monitored as samples undergo dark and light aging.

The benefits of this project to the conservation community and thus indirectly to the herbarium community are that it will:

- add to the body of fundamental knowledge on VAE emulsions,
- allow conservators and herbarium staff to know with more certainty what effect modifiers have on VAE emulsions and whether they are tolerable or detrimental, and
- provide conservators and herbarium staff with the ability to formulate emulsions tailored to their own needs.

METHYL CELLULOSE

Methyl cellulose has been used for mounting herbarium specimens. Although CCI has not carried out in-depth research on methyl cellulose, Feller and Wilt (1990) provide both a good review of cellulose ethers and their latest research results. According to this study, methyl cellulose is one of the most stable cellulose ethers. Nevertheless, methyl cellulose has poor adhesive qualities (e.g., tack and adhesive strength). If methyl cellulose was used in the standard plant mounting technique (i.e., drizzling adhesive

TABLE 3. Modifiers chosen to be studied in the VAE copolymer emulsion adhesive project.

Modifier	Chemical name	Commercial name	Supplier
Plasticizers	dibutyl phthalate diotridecyl phthalate polyethylene glycol dibenzoate	dibutyl phthalate Hatco 2922 Benzoflex® P-200	Sigma Chemical Co. Hatco Corp. Velsicol Chemical Co.
Solvents	toluene methanol diacetone alcohol	toluene methanol diacetone alcohol	Caledon Caledon Fisher Scientific
Wetting agents	di-(2-ethylhexyl) sulphosuccinate acetylenic glycol	Triton® GR-5M (anionic) Surynol® 104PA (non-ionic)	Union Carbide Air Products
Thickeners	methyl cellulose starch fumed silica polyethylene oxide acrylic acid salt	Methocel™ A4M Aytex P, wheat starch Cab-O-Sil® M-5 Polyox® WSRN 750 Acrysol® GS	Dow Chemical Canada Talas Cabot Corp. Union Carbide Rohm & Haas Canada Inc.
Fillers	calcium carbonate talc china clay	calcium carbonate talc Kaolin K-2	Sigma Chemical Co. Sigma Chemical Co. Fisher Scientific
Freeze-thaw stabilizer	ethylene glycol	ethylene glycol	Sigma Chemical Co.
Humectant	glycerin	glycerin	Fisher Scientific

over the plant in strategic spots), possible detachment could be expected due to these poor adhesive qualities. This is the reason why methyl cellulose is frequently used in conjunction with other adhesives (e.g., PVAC emulsions). Some herbaria immerse the whole plant in a methyl cellulose solution and then transfer it to the mounting paper. This method would certainly increase the surface area of the bond and consequently the total bond strength, but other concerns come to mind. For example, how removable is the adhesive, is the ethical conservation practice of "less is best" being followed, and is the chemical integrity of the specimen at risk? These are concerns that need to be considered.

MYLAR® TAPE ALTERNATIVE
PLANT MOUNTING SUGGESTION

CCI has been experimenting with an alternative plant mounting technique which avoids direct contact between the adhesive and the plant. It consists of a Mylar® strip whose ends have been coated with an acrylic pressure-sensitive emulsion adhesive, Rhoplex® N-619 (Fig. 2). The centre of the strip, which contacts the plant, is not coated with the adhesive. The strips can be made in any size or shape. Mylar® is a stable material and Rhoplex® N-619 adhesive has performed well in tests conducted by CCI (Down et al., 1996). The beauty of this technique is that if it is necessary to remount the plant, the strips are merely cut in the centre and the plant is easily removed.

Some preliminary testing of these strips was done at CCI. The strips were attached to paper and put in a dry oven at 80°C. How this oven aging relates to museum aging is not known at this time. After 20 months, the adhesive was still holding but the strips were easier to peel than strips that were left in the dark at ambient conditions. Also, slight staining of the paper from the adhesive occurred in the oven-aged strips.

Strips were also given to a few herbarium preparators for experimentation. Generally, they found that strips made with medium-weight Mylar® worked better than strips made with a thinner Mylar®. They also found that the strips did not stick to older, dusty mounting paper: this paper would have to be cleaned before applying the strips.

FIGURE 2. Illustration of Mylar® strip alternative. Strip width, strip length, and centre space width can be adjusted to individual need of specimen.

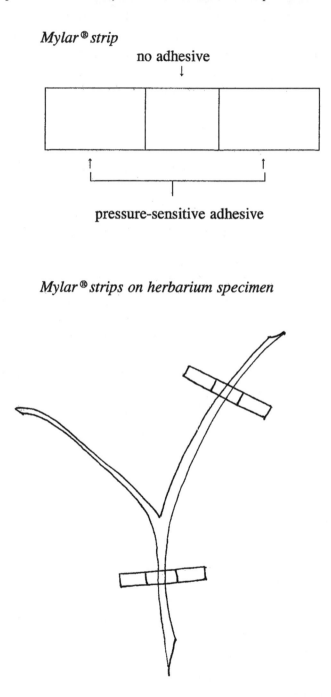

Certainly, more experimentation with this method and its long-term stability is needed. If proven suitable, the benefit is that it avoids adhesive contact with the plant which makes the treatment highly reversible and also helps to guard the genetic/chemical integrity of the plant (e.g., its potential for use in future DNA studies).

LABELS FOR HERBARIUM COLLECTIONS

CCI has examined two pre-gummed label products that could be used for natural history collections. Although it has come to our attention recently that changes have occurred to both these labels and they may no longer be available, the results of CCI's examination will be discussed because they show what is expected of a label.

In 1991, the Royal Ontario Museum asked CCI to examine Jac's paper label 10080-DC01 in the context of labelling microscope specimen slides (Down, 1991). This label was a pressure-sensitive type of label. The adhesive on the label was identified as a butyl acrylate. Other similar butyl acrylate adhesives tested by CCI (Down et al., 1996) were found to be neutral, flexible, non-yellowing, and to release no harmful volatiles — all good qualities. When tested, the extracted pH of the adhesive on the Jac's paper label was found to be in the neutral range at 6.65. The paper in the label contained no lignin or alum but starch was present. The surface pH of the paper was found to be 6.1. All of these results indicated that the Jac's paper label 10080-DC01 would have been suitable for the intended use.

The second label that CCI examined came from the Smithsonian Institution. It was a pre-gummed paper label made by Kanzaki Specialty Papers called Convertech. The adhesive on the label was identified as vinyl acetate/vinyl propionate (Moffatt, 1989). The surface pH of the adhesive side of the label was found to be 6.41 while the paper side of the label was found to be 6.8. Both pH values are in the neutral range. The paper in the label contained no lignin but starch was present. The paper contained some alum but not enough to be a problem. This label was aged in an oven for three weeks at 90°C and 50% RH. The surface pH of the paper side of the label, the surface pH of the adhesive side of the label, and the extracted pH of the paper and adhesive together were monitored after each week. The results indicate that the pH fell after aging but was still in the neutral range (Fig. 3). All of these results indicated that this label product would have been very promising for use in herbarium collections.

FIGURE 3. The effect of aging Convertech, a pre-gummed annotation label, for three weeks at 90°C and 50% RH on the surface pH of the paper side of the label, on the surface pH of the adhesive side of the label, and on the extracted pH of the paper and adhesive together.

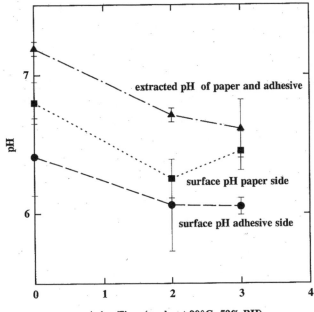

GLUE STICKS

Glue sticks are often used to attach labels to herbarium mounting paper. Analysis of three types of glue sticks determined that all were composed of poly(vinyl pyrrolidone) and had an alkaline pH of 9 (Williams, 1982 a, b, c). These analyses were performed in 1982 and should be re-checked because new glue sticks may have different compositions.

The concern with using glue sticks to glue labels to herbarium mounting paper comes from observations made at CCI on labels that had been glued 10 years earlier using a glue stick. These labels now peel extremely easily. As well, photographs glued at the same time as these labels lost their gloss in areas where the adhesive was on the reverse.

For these reasons, CCI does not recommend glue sticks for permanent adhesive solutions such as gluing labels to herbarium mounting paper. Glue sticks should only be viewed as a temporary gluing solution.

CONCLUSIONS

The perfect adhesive for herbarium collections is still elusive. Therefore, herbarium staff should always consider conservation criteria when choosing an adhesive for mounting specimens. Conservation research on adhesives can help to assist herbarium staff in their adhesive choices but there is still much research to be done. Certainly, the question of label adhesives still needs to be addressed and suitable available labels found. So too the use of acrylic adhesives for plant mounting needs to be further explored. The continued use of linen tapes for strapping plants would seem to be a good option since it avoids excessive adhesive on the plant and the starch and gelatin adhesives on these tapes are considered to be fairly stable. The search for a better emulsion adhesive continues. While Mowilith® DMC2 has relatively good aging properties, it may require more study on decreasing tack before it can be accepted whole-heartedly for mounting plants, if indeed it can be more easily obtained. Similarly, although Weldbond® was the best hardware-store-variety PVAC emulsion tested, it is too acidic, needs buffering, and requires further study on the effect of buffering. Perhaps the new VAE research will help in this respect. Herbarium staff will find it very beneficial to follow future conservation developments in all these areas.

ACKNOWLEDGEMENTS

Sincere thanks are extended to Mike Shchepanek and Micheline Bouchard in the CMN Herbaria for their assistance in helping to prepare this presentation. Also, thanks to John Pinder-Moss at the Royal British Columbia Museum, Botany Division, for his assessment of the Mylar® strips.

Literature Cited

Archer, W. A. 1950. New plastic aid in mounting herbarium specimens. Rhodora, Journal of the New England Botanical Club 52:624:298-299.

Down, J. L. 1991. Evaluation of Jac's paper label 10080-DC01. CCI Environment and Deterioration Research Report No. 1807.

Down, J. L. 1995a. Analysis of adhesive on linen tape from the Canadian Museum of Nature Herbaria. CCI Environment and Deterioration Research Report No. 2057.

Down, J. L. 1995b. Adhesive projects at the Canadian Conservation Institute. Pp. 4-12 in M. M. Wright and J. H. Townsend, editors. *Resins Ancient and Modern*. The Scottish Society for Conservation and Restoration, Edinburgh.

Down, J. L., M. A. MacDonald, J. Tétreault, and R. S. Williams. 1996. Adhesive testing at the Canadian Conservation Institute: an evaluation of selected poly(vinyl acetate) and acrylic adhesives. Studies in Conservation 41:1.

Feller, R. L. and M. Wilt. 1990. *Evaluation of Cellulose Ethers for Conservation*. Research in Conservation 3. The Getty Conservation Institute, Marina Del Ray, California, 161 pp.

Laver, M. E. 1975. Dennison Gummed Linen Tape. CCI Analytical Research Services Report No. 1330.

Moffatt, E. A. 1979. Holland Gummed Tape. CCI Analytical Research Services Report No. 1599.

Moffatt, E. A. 1982. Holland Gummed Tape. CCI Analytical Research Services Report No. 1844.11.

Moffatt, E. A. 1989. Analysis of the Adhesive on a Pre-gummed Paper from the National Herbarium. CCI Analytical Research Services Report No. 2795.

Moffatt, E. A. and G. S. Young. 1985. University Products Linen Gummed Tape. CCI Analytical Research Services Report No. 2343.2.

Williams, R. S. 1982a. Dennison Glue Stick. CCI Analytical Research Services Report No. 2017.

Williams, R. S. 1982b. Pritt Glue Stick. CCI Analytical Research Services Report No. 2011.

Williams, R. S.1982c. UHU Stic Glue Stick. CCI Analytical Research Services Report No. 2012.

APPENDIX 1
MATERIALS AND SUPPLIERS

Acrysol® GS, Rhoplex® N-619: Rohm & Haas Canada Inc., 2 Manse Rd., West Hill, ON M1E 3T9, Canada. Tel: (416)284-4711. Fax: (416)284-2982. Available from: AACS International Gilder's Supply Ltd., 12-1541 Startop Rd., Ottawa, ON K1B 5P2, Canada. Tel: (613) 744-0945. Fax: (613) 744-0949.

Airflex® 320, Airflex® 465, Surfynol® 104 PA: Air Products and Chemicals Inc., PO Box 25764, Lehigh Valley, PA 18002-5764, USA. Tel: (610)481-6799; (800)345-3148. Fax: (610)481-4381.

Aytex P (Wheat Starch), Jade No. 403, R-2258: Talas, Division of Technical Library Service Inc., 568 Broadway, New York, NY 10012, USA. Tel:(212) 219-0770. Fax: (212)219-0735.

Benzoflex® P-200: Velsicol Chemical Corp., 10400 W Higgins Rd., Suite 600, Rosemont, IL 60018, USA. Tel: (708) 298-9000, (800)843-7759. Fax: (708)298-9014.

Bondfast™: Lepage's Ltd., 50 West Dr., Bramalea, Ontario L6T 2J4, Canada. Available from Beaver Lumber Co. Ltd., 1412 Star Top Rd., Gloucester, ON K1B 4V7, Canada. Tel: (613)741-5052.

Bulldog Grip® 20 Minute Resin: Canadian Adhesives Ltd., 81 Kelfield Street, Unit 7, Rexdale, ON M9W 5A3, Canada. Available from M. Zagerman & Co. Ltd., 1630 Star Top Road, Gloucester, ON K1B 3W6, Canada. Tel: (613) 741-5990.

Bulldog Grip® Resin 2311: Canadian Adhesives Ltd., 81 Kelfield St., Unit 7, Rexdale, ON M9W 5A3, Canada. Tel: (416)241-1151.

Cab-O-Sil® M-5: Cabot Corp., PO Box188, Tuscola, IL 61953-0188, USA. Tel: (217)253-3370, (800)222-3370. Fax: (217)253-4334.

Calcium Carbonate, Dibutyl Phthalate, Ethylene Glycol, Talc: Sigma Chemical Co., PO Box 14508, St Louis, MO 63178-9916, USA. Tel: (800)325-3010. Fax: (800)325-5052.

Convertech: Kanzaki Specialty Papers, 20 Cummings St., Ware, MA 01082, USA. Tel: (413)967-6204. Springfield Sales Office. Tel: (413)736-3216.

Diacetone Alcohol, Glycerin, Kaolin K-2: Fisher Scientific, 112 Colonnade Rd., Nepean, ON K2E 7L6, Canada. Tel: (613)226-8874. Fax: (613)226-8639.

Dur-O-Set® E-150: National Starch and Chemical Co., 10 Finderne Ave., PO Box 6500, Bridgewater, NJ 08807-3300, USA. Tel: (908)685-5000. Also available from: Nacan Products Ltd., 60 West Dr., Brampton, ON L6T 4W7, Canada. Tel: (905)454-4466. Fax: (604)596-0441.

Elmer's® Glue-All: Borden Chemical Canada, 595 Coronation Dr., West Hill, ON M1E 4R9, Canada. Available from Builders' Warehouse, 3636 Innes Rd., Orleans, ON K1C 1T1, Canada. Tel: (613) 824-2702.

Elvace® 40705, Elvace® 40709, Elvace® 40724: Reichold Chemicals Inc., PO Box 13582, Research Triangle Park, NC 27709-3582, USA. Tel: (919)990-7500. Fax: (919)990-7711.

Gaylord Magic-Mend: Gaylord Library Supplies and Equipment, PO Box 4901, Syracuse, NY 13221-4901, USA. Tel: (315)457-5070.

Hatco 2922: Hatco Corp., 1020 King George Post Rd., Fords, NJ 08863, USA. Tel: (908)738-1000. Fax: (908)738-9385.

Jac's Paper Label 10080-DC01: Jacpapier Inc., 980 boul. Roche, Vaudreuil, PQ J7V 8P5, Canada. Tel: (514)455-7971.

Lascaux 498HV: Conservation Materials Ltd., 1165 Marietta Way, PO Box 2884, Sparks, NV 89431, USA. Tel: (702)331-0582. Fax: (702)331-0588.

Methanol, Toluene: Caledon, Laboratories Ltd., 400 Armstrong Ave., Georgetown, ON L7G 4R9, Canada. Tel: (416)456-0226, (800)668-3230. Fax: (416)877-6666.

Methocel™ A4M: Dow Chemical Canada, 1086 Modeland Rd., PO Box 1012, Sarnia, ON N7T 7K7, Canada. Tel: (519)339-3131.

Mowilith® DMC2: Was available from - Carr McLean, 461 Horner Ave., Toronto, ON M8W 4X2, Canada. Tel: (416)252-3371. Fax: (416)252-9203. Also sold as Appretan MB Extra or Nacan 685773. Manufactured by Nacan Canada for Hoechst Canada. Samples have been obtained recently from Hoechst Canada. Tel: (514)333-3597.

Polyox® WSRN 750, Triton® GR-5M: Union Carbide Canada Inc., 1210 Sheppard Ave. E, Suite 210, Box 38, Willowdale, ON M2K 1E3, Canada. Tel: (416)490-0052. Fax: (416)490-0051.

Weldbond®: Frank T. Ross & Sons (1962) Ltd., PO Box 248, West Hill, ON M1E 4R5, Canada. Available from Beaver Lumber Co. Ltd., 1412 Star Top Rd., Gloucester, ON K1B 4V7, Canada. Tel: (613)741-5052.

Chapter 12

Testing Dry Documentation Media for Permanent Hard-Copy Collection Records

Stephen L. Williams and R. Richard Monk

Abstract.—As collection workers become increasingly cognizant of the need to ensure long-term preservation of permanent records, there is an ongoing challenge to select appropriate documentation media. Ideally, this selection should be based on a thorough analytical examination of the available products. However, few institutions have the ability and equipment necessary for such examinations. Furthermore, products are continually being modified or replaced by manufacturers. Given these realities, the identification and avoidance of an undesirable product can be as important as identifying the best product. There are simple tests to assess lightfastness, solubility, and adherence of documentation media. Any product that does not perform satisfactorily in these tests might be expected to perform poorly over time.

INTRODUCTION

In collection settings, the accumulation and preservation of hard-copy documentation is an important aspect of collection operations. Such documentation is associated with specimens, field records, permits, collection record-keeping, archives, surveys, reports, and research. In most instances, this documentation is maintained because of its long-term relevance and because of its understood value for fulfilling responsibilities to the institution, to society, and to the legal systems. Because hard-copy documentation is vital to many collection operations, it is important that the integrity of the information not be compromised by the use of inappropriate documentation media and substrates.

There are numerous forms of media used for hard-copy documentation including pencil, various kinds of inks, and others that are printed, stamped, typed, or photocopied. These media can be applied to a variety of substrates including paper, plastic film, and textiles. Furthermore, technological advances continually result in the creation of new products (e.g., laser printers, photocopiers, and FAX machines), new applications (e.g., marking cryogenic materials and bar-coding), and product modifications (Williams and Hawks, 1988). It is the responsibility of the collection staff to discern the quality of various products before they are used for permanent collection records. While it is desirable to select superior products, it can be more important to avoid those that are known to be problematic. For instance, knowing that a product will fade should prompt one to seek a more lightfast product.

Depending on the nature of a documentation medium, several experiments can be designed to test and compare product qualities (US Department of Commerce, 1920). For instance, it is easy to test a liquid medium for pH, corrosiveness, total solids, drying time, and fluidity (Williams and Hawks, 1986). However, most documentation media are encountered in a dry form, thus restricting possible testing methods. There are some simple tests that can assess dry documentation media for properties of lightfastness, solubility, and adherence. While they may not be comprehensive enough to endorse a product, they will help to separate problematic products from those that may be suitable for collections use. The following describes basic procedures required to perform these tests.

SAMPLE PREPARATION

The preparation of samples is standard for most tests. In all cases, the documentation medium must be applied to a stable, non-reactive substrate, so that interpretation of test results will be based on the medium alone and not unrelated interactions involving the substrate. Similarly, due consideration should be given to the ways in which substrate characteristics may alter the performance of the medium. For instance, a plastic film is less appropriate for testing a penetrating fluid ink but is more appropriate for a powdered toner that is mechanically or electrostatically applied. A paper substrate is typically used for most hard-copy documentation because of its absorbance, contrast with most media, availability, cost, and durability (Hawks and Williams, 1986). For testing purposes, a good quality white paper that has 100% cotton fiber content, medium weight, regular finish, and is acid-free is recommended. If other substrates, such as plastic films or textiles, are to be used with the medium, they should be incorporated in the tests.

For most tests, the documentation media are applied to the substrate in a series of broad lines. If more than one medium is to be tested, it is recommended that they be tested simultaneously, and that each medium be applied parallel to the others on the substrate. This will result in a series of separated rows of media to be tested. Test strips are made by cutting the substrate perpendicular to the rows of media so that each strip includes a sample of each medium to be tested. Strips should be 5 to 10 centimeters wide and at least 20 centimeters long. This procedure will allow an easy comparison of the treated (test) and untreated (control) areas, while ensuring consistency of application, testing, and assessment of results.

LIGHTFASTNESS

Documentation media must resist fading and any photochemical change that might compromise the integrity of the substrate. There are standard testing methods for evaluating lightfastness of printed matter which make use of either xenon-arc lamps, carbon-arc lamps, fluorescent light fixtures (fluorescent-light method) or daylight behind window glass (window-glass method; ASTM, 1994a). Modified versions of the latter two methods are simple to implement and require a minimum of resources. Thus, either of these methods can be conducted with comparative ease.

An important consideration when testing for lightfastness is to provide reasonable and stable climatic conditions for the duration of the testing. Direct or indirect contact of the test strips with surfaces that may influence temperature and moisture levels must be avoided. Providing ventilation around the test area will minimize the risk of either the medium or the substrate being affected by fluctuations in temperature or moisture level.

The amount of light that the samples are exposed to should be measured. This is accomplished with a blue-wool fade card, available from conservation supply distributors. The ISO blue-wool fade card is a recognized monitoring standard of the International Standards Organization (Feller and Johnston-Feller, 1978, 1979, 1981). It consists of eight pieces of wool cloth colored with different blue dyes that are progressively faded by increasing amounts of light. A single card can be cut into strips and used for test and control purposes.

To test for lightfastness, both covered and uncovered test strips and blue-wool fade card strips (Fig.1) are exposed to a source of

FIGURE 1. Lightfastness is tested by exposing documentation media to ultra-violet light. An ISO blue-wool fade card is used to measure light exposure. Covering the lower half of the media test strip and a strip of the fade card with aluminum foil provides unexposed controls. Sample B illustrates fading.

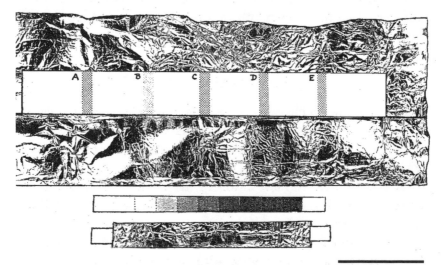

5 cm

ultraviolet radiation. An unobstructed window that is fully exposed to sunlight during daylight hours should be used when employing the window-glass method. An opaque, non-reactive material, like aluminum foil, should be used to cover the strips of blue wool fade card and parts of the test strips that are to be protected from light exposure.

The lightfastness test should be continued at least until the sixth blue-wool sample is faded. This may require at least six months of continuous exposure (Wood and Williams, 1993). Materials capable of withstanding this amount of exposure without change can be expected to have an "intended useful lifetime greater than 100 years" (Feller and Johnston-Feller, 1981). Any documentation medium that fails this test (e.g., Fig. 1, sample B) should not be used for permanent collection documentation. Tests for lightfastness of documentation media have been reported by Lafontaine (1978), Grant (1985), Williams and Hawks (1986), Townsend (1986, 1990), Lavrencic (1989), Ramsay and Thomson (1990), Maraval and Flieder (1992, 1993), and Wood and Williams (1993).

SOLUBILITY

Documentation media may come in contact with fluids in collection storage or cleaning treatments, or inadvertently through accidental spills, condensation, grease or resin migration, or floods. Thus, it is important to know which fluids may compromise the integrity of the documentation medium in question. Solubility testing should always include water, but may also include other fluids that might be encountered in the collection such as formalin, ethanol, isopropanol, glycerin, and ammonia. Knowing which fluids may act as solvents is important both for protecting documentation and for possibly identifying components of the documentation medium (ASTM, 1994b).

The technique used to test for the solubility of documentation media is similar to that used in paper chromatography. The ends of the test strip are clipped together to form a cylinder (Fig. 2). The bottom portion of the paper cylinder is then lowered into the test fluid. The test fluid will be absorbed by the paper and eventually migrate into contact with the media samples. If the fluid reacts with a medium (e.g., sample A of Fig. 2), the effects will be evident when the wet and dry parts of the paper strip are compared. It is not unusual for components of a given documentation medium to become obviously separated in the wet area above the fluid line. If the medium is strongly affected by the solvent, it may dissolve. Any documentation medium that fails important solubility tests, especially water, should not be used for permanent collection records. Tests for solubility of documentation media have been reported by Williams and Hawks (1986), Lavrencic (1989), Horie and Barry (1990), Townsend (1990), and Wood and Williams (1993).

FIGURE 2. Solubility is tested by exposing documentation media to different test fluids. The effect of the fluid on the media is monitored as the fluid migrates up the paper substrate. Sample A illustrates dissolving.

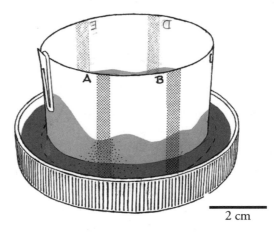

2 cm

ADHERENCE

Documentation media attach to substrates in one of two ways depending on the nature of both the medium and the substrate. In some cases, such as fluid inks on paper, the ink medium is initially in a liquid state and is able to penetrate the paper substrate before drying. The penetration affords relative permanence because it provides resistance to mechanical degradation of the medium. Alternatively, with dry inks, the medium is attached to the substrate surface by adhesive forces created by mechanisms such as pressure (e.g., print from plastic typewriter ribbon), heat-fusion, or electrostatic charge (e.g., toner on polyester plastic film). Depending on the quality of the adhesion, the medium may eventually become physically separated from the substrate by abrasion or flaking resulting in a loss of information. Because such loss of information is not acceptable for permanent records, it is important to test adherence of documentation media with both porous and non-porous substrates. A useful adhesion test can be conducted with either cotton swabs or with clear plastic adhesive tape (e.g., 3M Scotch Magic Tape™).

For the swab test, a clean cotton swab is placed in contact with the substrate and firmly wiped across the documentation medium (Fig. 3a). This process may need to be repeated three to five times. If no change is observed in either the medium, the substrate, or the swab, the medium may have desirable properties for resisting abrasion. In contrast, removal of the medium onto the swab or smearing of the medium across the substrate or across the swab may indicate that the medium exhibits undesirable properties. It is possible that excess medium may be removed, particularly on the first stroke, potentially leading to a misinterpretation of results. If there is a doubt about the results, retest the medium with a clean swab.

The adhesion results can be further verified by using clear plastic adhesive tape. For this test, a strip of tape is placed across the documentation medium on the test strip. The tape is first gently pressed onto the substrate to ensure complete adhesion and then carefully removed (Fig. 3b). Any removal of the documentation medium from the test strip may indicate a tendency for the medium to flake away from the substrate as it ages or is altered by environmental conditions. With either test, it is advisable to examine the substrate, the medium, and the swab or tape surfaces under magnification to determine whether the testing affected only the medium, only the substrate, or both.

Adherence testing need not be restricted to the techniques described. For instance, a comparison of the effects of different plastic and rubber erasers to documentation media may also provide

FIGURE 3. Adherence is tested by evaluating the resistance of documentation media to physical treatments. (a) Resistance to abrasion is tested by wiping a swab across the media. Samples D and E illustrate media that have been removed from the substrate by the swabs. (b) Resistance to flaking is tested by gently pressing clear plastic adhesive tape across the media and then carefully removing it. Samples A and C illustrate media that have been removed from the substrate by the tape.

(a)

(b)

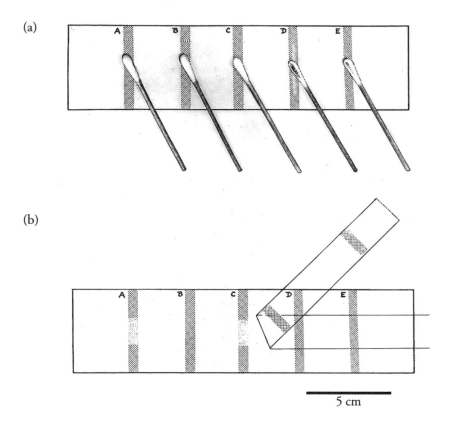

5 cm

useful information. Similarly the potential for documentation media to be affected by plasticizers when in contact with some plastic surfaces (e.g., poly(vinyl chloride) film) warrants further investigation. Unlike the tests described for lightfastness and solubility, adherence testing incorporates variables that are difficult to eliminate or standardize, such as pressure. Nevertheless, any documentation medium that fails the described adherence tests should be carefully scrutinized before it is used for permanent documents. Results of adherence testing are reported in Wood and Williams (1993).

CONCLUSION

There are several factors that make the choice of documentation product difficult. Most products may perform well for a limited number of criteria, but rarely for a majority of the criteria. For instance, one product may be resistant to water but not to alcohol. Also, it is a challenge to sift through the growing number of products to find those suitable for the wide range of documentation needs that exist in most collections. Finally, while it is both expected and desired that there be some consistency in product formulations, in reality changes occur with technological advances, availability and cost of product ingredients, and legislation for environmental protection (Grant, 1985; Williams and Hawks, 1986; Williams and Hawks, 1988; Wood and Williams, 1993). Thus, until manufacturers more actively address all of the archival requirements of documentation products, it is not likely that any single medium will be perfect for all situations. These issues make it both appropriate and necessary to identify testing methods and to periodically examine documentation media used for permanent documents.

LITERATURE CITED

[ASTM] AMERICAN SOCIETY FOR TESTING AND MATERIALS. 1994a. ASTM 3424 — Standard test method for evaluating the lightfastness and weatherability of printed matter. Annual Book of ASTM Standards (Paints, Related Coatings, and Aromatics) 6(2):173–178.

_____. 1994b. ASTM E 1422 — Standard guide for test methods for forensic writing ink comparison. Annual Book of ASTM Standards (General Methods and Instrumentation) 14(2):854–860.

FELLER, R. L. AND R. M. JOHNSTON-FELLER. 1978. Use of the International Standards Organization's Blue-wool Standard for exposure to light. I. Use as an integrating light monitor for illumination under museum conditions. Pp. 73–80 in *Reprint of American Institution of Conservation for Historical and Artistic Works* (6th Annual Meeting), Austin, Texas.

_____. 1979. Use of the International Standards for exposure to light. II. Instrumental measurement of fading. Pp. 1–7 in *Preprint American Institute of Conservation for Historical and Artistic Works* (7th Annual Meeting), Toronto.

_____. 1981. Continued investigation involving the ISO Blue-wool Standards of exposure. ICOM Committee for Conservation. Ottawa Paper 81(18):1–7.

Grant, T. 1985. Lightfastness of alternative colourants for textiles. Master's Thesis, Queen's University, Kingston, Ontario.

Hawks, C. A. and S. L. Williams. 1986. Care of specimen labels in vertebrate research collections. Pp. 105–108 in J. Waddington and D. M. Rudkin, editors. *Proceedings of the 1985 Workshop on Care and Maintenance of Natural History Collections*. Life Sciences Miscellaneous Publications, Royal Ontario Museum, Toronto.

Horie, V. and J. Barry. 1990. Solvent resistance of marking inks. Conservation News 41:11–12.

Lafontaine, R. H. 1978. The lightfastness of felt-tip pens. Journal of the International Institute for Conservation — Canadian Group 4(1):9–16.

Lavrencic, T. 1989. Ink permanence testing. Newsletter of Australian Institute for Conservation of Cultural Materials 30:11.

Maraval, M. and F. Flieder. 1992. Etude de la stabilité des encres d'imprimerie. Nouvelles de l'ARSAG et du Groupe documents graphiques du Comité de conservation de l'ICOM 8:20.

_____. 1993. The stability of printing inks. Restaurator 14(3):141–171.

Ramsay, P. and D. Thomson. 1990. Pens for museum documentation. Conservation News 43:12–14.

Townsend, J. H. 1986. Lightfastness of marker pens. Conservation News 30:6–7.

_____. 1990. Labelling and marker pens. Conservation News 42:8–10.

US Department of Commerce. 1920. Inks — their composition, manufacture, and methods of testing. Circular of the National Bureau of Standards 95:1–24.

Williams, S. L. and C. A. Hawks. 1986. Inks for documentation in vertebrate research collections. Curator 29(2):93–108.

_____. 1988. A note on "Inks..." SPNHC Newsletter 2(1):1.

Wood, R. M. and S. L. Williams. 1993. An evaluation of disposable pens for permanent museum records. Curator 36(3):189–200.

CHAPTER 13

Plastic Materials Used in the Herbarium

JULIA FENN

Abstract.—Plastics used for the storage and mounting of plant material (envelopes, boxes, petri dishes, gaskets, and seals) may disintegrate or may damage the specimens if the polymer has unsuitable chemical or physical properties. Some of the strengths and weaknesses of different museum plastics are discussed in the context of their use in herbarium collections so that harmful plastics or pointless expense can be avoided. Some simple testing guidelines are identified to assist in the determination of polymer types.

INTRODUCTION

It has been recognised for many years that museum collections can be damaged by corrosive vapours given off by the storage materials intended to protect them (Oddy, 1973; Blackshaw and Daniels, 1978). Certain plastics, for example cellulose acetate, cellulose nitrate, poly(vinyl chloride), and vulcanised rubbers, can be even more harmful than notorious villains such as oak or plywood. Now that museums are being affected by severe financial constraints, it is vital to choose safe storage and display materials because there is neither the time nor the money available for frequent replacement of storage materials or disintegrating specimens (even assuming that the specimens are replaceable). There are several publications that discuss testing methods (Oddy, 1975; Blackshaw and Daniels, 1979; Tétreault, 1993; Zhang et al., 1994) and recommended materials (Blackshaw and Daniels, 1978; Hopwood, 1979; Fenn, 1990).

Of the many kinds of plastics available only a few are considered sufficiently non-corrosive to use with artefacts and specimens, namely: polyethylene, polypropylene, polystyrene, poly(methyl methacrylate), polycarbonate, poly(ethylene terephthalate) and poly(tetra fluoroethylene). Each of these plastics has its own particular strengths and weaknesses and it is useful to know their individual properties so that the most cost effective plastic can be chosen for any particular application.

One especially important property is resistance to biocides. Many botanical specimens are heavily contaminated with one or more of the aromatic biocides such as paradichlorobenzene, naphthalene, camphor, thymol, and dichlorvos. In high concentrations these vapours attack many plastics and adhesives causing shrinkage, stickiness, deformation, and discoloration (Dawson, 1984). Some biocides appear to act in synergism so that the damage caused by combinations of different insect repellant vapours may be unexpectedly severe even at comparatively low concentrations (Sakata and Yamada, 1979; Waddington and Fenn, 1988). Many of the aromatic biocides are also a health hazard (Rossol, 1995) and if staff are repeatedly exposed, the plastics may not be the only casualties. Threshold limit values for some of the common moth repellant insecticides are given in Appendix A.

PROPERTIES OF "SAFE" MUSEUM PLASTICS

Polyethylene

Polyethylene (PE) is a translucent polyolefin available as sheets and bags (e.g., polythene, Tyvek™, Ziploc™ bags) or containers (e.g., Tupperware™). It is also available as foams of varying density (e.g., Ethafoam™, Sentinel™ foam) or the more expensive cross-linked foams (e.g., Minicel™, Volara™, Plastazote™). Radiation cross-linking is preferable to chemical cross-linking because there is less danger of harmful residues. PE is resistant to most solvents and biocides, but swells and distorts after long-term exposure to hydrocarbon solvent vapours or direct contact with oils or waxes. Consequently, PE and other polyolefins should not be used with botanical wax models. Ordinary low-density PE is subject to stress cracking at low temperatures. Special freezer-resistant grades have been developed for the food industry and these kinds are suitable for freeze-drying specimens or for freezing to control insect infestations. The great advantage of polyethylene containers is their impact resistance and the fact that they can often be sealed to prevent the entry of water or insects. However they are not transparent except in very thin section, which can be a disadvantage.

Polypropylene

Polypropylene (PP) is a translucent polyolefin with properties and museum uses similar to PE. It is usually slightly more expensive but the relative cost of polyolefins has tended to fluctuate more than other plastics. One of its most common forms is an opaque, corrugated sheet (e.g., Corotex™, Coroplast™) used to make storage boxes and supports for herbarium sheets. PE and PP boxes often have hinges made from a thin, biaxially-oriented polypropylene membrane (e.g., Kodak™ slide boxes). Experiments at the Royal Ontario Museum (ROM) have shown that long-term contact with organic solvents and oils can relax the orientation of the molecules, destroying the fold resistance of the hinges. A crystal clear polypropylene for the food packing industry is now available but it is expensive and as yet it has not been tested for museum use.

Polystyrene

Polystyrene (PS) is one of the cheapest and most common plastics. Crystal PS is completely transparent but has poor resistance to light, impact, or abrasion. Vitrines and specimen boxes made from crystal polystyrene often become weak and cloudy as tiny stress cracks (crazing or crizzling) develop after exposure to solvent and biocide vapours. However, in dimly lit storage conditions with good air quality, it has a good life expectancy (Byers, 1993).

Acetic and formic acids are partial solvents for PS and recent experiments have shown that the plastic can become corrosive if it has been exposed to plywood or any other source of these volatile acid vapours. This contamination can occur at levels which are too low to affect the appearance or the odour of the plastic (Fenn, 1995). PS has excellent resistance to freezing temperatures and is used commercially in refrigeration units. Expanded PS in the form of opaque, white beadboard or S-shaped chips (e.g., Styrofoam™) is a useful packing material which does not appear to absorb organic acids to the same extent as crystal PS. It is lightweight and less abrasive than most expanded polyolefins, but not as hardy. A real disadvantage is its tendency to shrink if crated for long periods with specimens contaminated with vapour phase biocides such as naphthalene and paradichlorobenzene.

Poly(methyl methacrylate)

Poly(methyl methacrylate) (PMMA) is a hard, transparent acrylic often used for vitrines (e.g., Plexiglas™, Lucite™, Perspex™) and sometimes for specimen boxes. It has excellent resistance to oxidizing agents and oils, and has better impact resistance than crystal polystyrene. Like polystyrene it can become corrosive after exposure to acetic or formic acid vapours (Fenn, 1995) and it is similarly vulnerable to stress cracking caused by aromatic biocides and solvents, particularly alcohol and ammoniated cleaners. PMMA is also reputed to have poor resistance to glycerine preservatives. It has a high coefficient of expansion and as temperature rises, it expands more than most metals. This can cause cracking where inflexible fastenings or glue lines interfere with movement in mounts and display cases, especially during travelling exhibitions. The manufacturing method has a significant effect on the properties of sheet PMMA. *Extruded* PMMA (the most common form) expands much more in the direction of extrusion than at right angles to it. For construction purposes it is better to buy *pre-shrunk, extruded* PMMA as the molecular alignment has been relaxed with controlled heat treatments so that differential shrinkage is reduced. *Cast* PMMA, on the other hand, has little directional differentiation, has better impact resistance, and slightly better sag resistance, but it is more expensive, is often optically flawed, and is not readily available in large sizes.

Polycarbonate

Polycarbonate (PC) is a rigid, transparent polyester with a slightly grey hue. Its chief advantage over PMMA and PS is superb impact resistance so its main use in museums is as vitrines (e.g., Lexan™, Tuffak™). Thin, flexible PC foil can also be slipped over the glass in framed botanical prints to protect the glass; if the foil is thin enough it does not seriously affect the optical properties. Exposure to common solvents, especially alcohols and ammoniated glass cleaners, are reputed to destroy its impact resistance before there are any visible signs of stress cracking. PC is yet another of the plastics which can become corrosive after exposure to low levels of acetic or formic acid vapours (Fenn, 1995).

Poly(ethylene terephthalate)

Poly(ethylene terephthalate) (PET) is a polyester available as a flexible, transparent film (e.g., Mylar™, Melinex™), as woven fabric (e.g., Dacron™, Terylene™), or as unwoven felts and batting or fibrefill (e.g., Hollytex™, Reemay™). Some grades of polyester felt are bonded with potentially corrosive adhesives rather than with heat. To avoid this problem, buy only reputable brands whose manufacturers provide reliable manufacturing data on their products. PET has good dimensional stability and solvent resistance. Sealed bags and envelopes made from the film have low permeability to many fumigants and are not affected by aromatic biocides. Polyester laminated with metal foil (e.g., Marvelseal™ 360) is an effective vapour barrier against emissions from acidic woods and plywoods. Mylar™ D is the type normally recommended for museum collections but be wary of thick sheets which have very sharp edges and can gouge specimens and staff. Frosted or matte polyesters may have abrasive inclusions (e.g., Mylar™ EB-11). PET has few other problems apart from notch sensitivity (it tears easily) and weakness at the weld in heat-sealed polyester envelopes. As a result, envelopes are sometimes made using pressure sensitive adhesives which have a tendency to soften and creep.

PET is expensive, but because of its strength, it can be used as a very thin film or be laminated with another material such as PE to reduce its cost and its tendency to tear, and to improve the strength of heat-sealed bonds.

Poly(tetra fluoroethylene)

Poly(tetra fluoroethylene) (PTFE), usually known by the trade names Teflon™ or Fluon™, is a white fluoropolymer with a very high melting point and excellent chemical resistance to virtually everything except hot, concentrated alkalis and gamma irradiation. PTFE stays tough and flexible at temperatures well below freezing. Teflon™ tape, pricked with the identification number, is a useful method of tagging specimens undergoing treatments which destroy other labelling materials. Plumbers tape, available from hardware stores, is a comparatively inexpensive form of Teflon™ tape that can be used for this purpose. PTFE is also manufactured as an opaque, waterproof fabric (Gore-tex™) which allows most vapours (but not liquids) to permeate through tiny micropores. It has been used as protective clothing against corrosives and is useful for bagging specimens for fumigation, or enclosing sorbents such as silica gel during humidity control. PTFE is one of the most expensive plastics, which has limited its use in museums.

Polyurethane and silicone rubbers

Occasionally other polymers such as silicone rubbers and some grades of dense polyurethane rubbers have been found acceptable for museum use, providing that they have been manufactured without harmful additives. However, in general, they are not recommended. Polyurethane foams, in particular, often have a very short lifespan. It is very difficult to estimate the useful life expectancy of polyurethanes, or indeed any polymer, because accelerated aging tests have proved to be unreliable.

UNDESIRABLE PROPERTIES IN MUSEUM PLASTICS

Any plastics, even those recommended for museum use, can be damaging to the collections if they are badly made, or if harmful additives have been incorporated during manufacture. For example, recycled and biodegradable plastics cannot yet be recommended for museum use because the quality is still inconsistent. Recycled polyester and polyethylene sheet has sometimes been found to have splits, discoloured areas and abrasive inclusions of semi-melted granules. Biodegradable plastics are actually designed to have a shorter lifespan and, in addition, may have edible components such as starch to encourage insect and bacterial attack. This will not help the pest management program!

It is also better to avoid coloured plastics, with the exception of virgin polymers opacified with purified carbon black. Many colouring agents have an adverse effect on the aging properties of the plastic. According to one former manufacturer, dyes are sometimes used deliberately to disguise discolouration caused by poor manufacturing processes. Blue, black and green are the colours most frequently used for this purpose. Any black plastic intended for use with the collections should therefore be purchased from a reputable manufacturer who will guarantee the quality of the product. Black garbage bags from discount stores may be delightfully cheap, and excellent as garbage bags, but if museum staff use them as support materials in the storage room, they are likely to stain the collections.

As a general rule there is little advantage in using plastics which have special additives to alter fire resistance, static charge, or biocidal properties. These additives, while adding to the expense, are often short-lived or harmful to the specimens. For example, there may be some museums which are using "insect deterrent" vinyls intended for picnic tablecloths. It is probably not even advisable to picnic off them in the open air; in an enclosed storeroom they are definitely not a good idea.

IDENTIFICATION OF PLASTIC MATERIALS ALREADY IN USE IN HERBARIA

Many plastic materials with very different properties look very similar. When assessing plastic storage materials already in use, it can be difficult to know what they are and whether they are harmful enough to warrant the time and cost of replacing them.

It is also useful to be able to check new orders. If an expensive plastic has been ordered for a specific purpose, it can be very annoying to discover, at a later date, that a plastic of similar appearance but inferior performance, has been substituted. The Botany Department at the ROM specified that the seals on their new herbarium cases were to be made from dense silicone rubber foam. A cheaper, cross-linked polyethylene foam, with poor long-term resilience, was substituted without notification.

There are several simple techniques for identifying plastics when analytical facilities are not available, such as float tests, solvency tests, burn tests, and tear tests. An extremely useful sample kit with instructions is available from the Taylor Made Company (Taylor, 1988). Precise identification can be confused by additives and lamination (Coxon, 1993) but in most cases it is possible to eliminate harmful plastics without too much difficulty. For example, chlorinated plastics such as poly(vinyl chloride), which have the potential to release hydrochloric acid and oily plasticizers, can be identified by their reaction with copper wire to produce a green flame in the Beilstein test. Copies of this test and the diphenylamine spot test for cellulose nitrate are available from the Canadian Conservation Institute (CCI; Williams, 1988, 1993).

To simplify identification, the plastics have been organised according to their common use in herbaria.

Plastic Films used for Envelopes, Interleaving, and Encapsulation

The polymer most commonly recommended for interleaving or encapsulation is poly(ethylene terephthalate) (e.g., Mylar D™). It can be identified by its clarity coupled with its resistance to common solvents. (Test both sides in case it has been laminated.) Polyesters give a negative result to the Beilstein test (Williams, 1993).

Polyethylene and polypropylene are also commonly used because of their ready availability. They have the additional advantages of being resistant to tearing and easy to heat-seal. They can be distinguished from other plastic films by their ability to float in water to which a drop of detergent has been added.

Sheet plastics are sometimes manufactured by stretching the

polymer simultaneously in two directions, which reorients the molecules and greatly improves the fold strength. Biaxially-oriented plastics are more expensive and their excellent fold strength is not necessary for most museum purposes.

Before flexible transparent plastic sheeting was available, reconstituted cellulose (Cellophane™) was used and many elderly examples are still present in the collections. They are usually quite acidic and with age they become very brittle and yellow. Cellophane™ has very little moisture resistance so it was often coated with corrosive polymers such as cellulose nitrate or cellulose acetate. Cellulose nitrate-coated sheets should be removed immediately because degrading cellulose nitrate gives off very acidic, oxidizing nitrogen dioxide gas which can weaken both the specimen and the herbarium sheets to the point of disintegration. It is easy to check whether Cellophane™ is coated by touching it with a wet swab or a tiny drop of water. Naked Cellophane™ will swell locally to form a rippled spot within two to three minutes; coated Cellophane™ will either not absorb water at all, or will take longer than 30 minutes before there is any effect. All coated Cellophane™ is likely to be harmful, but the diphenylamine test can be used to distinguish cellulose nitrate from other plastic coatings (Williams, 1988). The test solution contains concentrated sulphuric acid so a tiny fragment of cellophane should be detached to avoid contaminating the herbarium sheets. Because of the corrosive properties of sulphuric acid, the testing should be conducted in a well-ventilated area with access to water.

Plastics used for Boxes, Vials, Jars, and Petri Dishes

Many collections are still using glass which can be distinguished from the plastic substitutes because it is much colder to the touch. Otherwise, most rigid, transparent plastic containers in the collection will be made from polystyrene, which comes in a variety of useful sizes. Crystal PS is a noisy plastic and a simple test is to tap it with a pencil or to drop it. A loud metallic clatter is characteristic of PS. It is useful to have control pieces of similar shape so the sound can be compared, but in most cases the noise is so distinctive that it is not necessary to follow up with solvent tests (Taylor, 1988).

Poly(methyl methacrylate) containers are useful for specialised boxes such as those with lenses incorporated into the lids. Otherwise, their better optical clarity, and resistance to impact and light degradation are not of particular value in normal storage conditions. PMMA's susceptibility to aggressive vapours is still likely to be within the concentrations found in many collections. These factors, together with its much lower resistance to freezing and cold storage, should be considered before selecting PMMA over PS for storage containers.

Polyethylene and polypropylene containers are distinguishable by their cloudy translucence and their low specific gravity which allows them to float in water.

Identification of petri dishes and specimen jars is difficult because they are available in a range of specialised laboratory plastics. Salespersons often encourage museum staff to buy better quality "scientific" plastics such as poly(methyl pentene), rather than cheaper non-specialised kinds made for common use. Such polymers seem to be perfectly safe for the collections, however, the properties that make them so expensive, such as resistance to repeated autoclaving or gamma irradiation, may be quite unnecessary for general storage purposes. If there is a need for plastics with specialised properties, it is always better to discuss it with the company chemists rather than the sales department who are often uninformed about the products they sell.

Lids, especially flexible lids, should also be tested to ensure that they have not been made from plasticised poly(vinyl chloride). Screw top lids on old jars are often made from phenol formaldehyde (Bakelite™) which is usually black, brittle, and very inflexible. Phenol formaldehyde is a thermoset so it will not melt if touched with a red hot wire. These are being superseded by polyolefin lids which are cheaper to produce and give a tighter fit. In damp environments, phenol formaldehydes can emit both carbolic and formic acids, but replacing the lids is not a high priority unless they are continually wetted with water or alcohol.

Herbarium Case Gasket Materials

Traditional materials for sealing herbarium case doors were woollen felts or vulcanised rubber seals, both of which release sulphur compounds and often lose their resilience so there is no longer a tight fit. A piece of polished silver placed adjacent to old gaskets will tarnish within a month if sulphur compounds are being emitted.

More recently, inexpensive, low density, open-celled polyurethane foams have been used. This material, which initially provides a resilient seal against dust and insects, rapidly flattens and eventually disintegrates into a useless powder with a high static charge which clings to the specimens. There are a few polyurethane foams of greater resistance and longevity, but those often emit oily plasticizers.

Poly(tetra fluoroethylene) (Teflon™, Gore-tex™) gaskets have excellent chemical stability, but they are very expensive and, in their current form, they do not have the necessary physical resilience to give a tight fit after the doors have been opened a few times.

Polyolefin seals, in the form of cross-linked foams, are available in a range of densities. One source is hardware stores where they are sold as weather-stripping. They can usually be distinguished from polyurethane weather-stripping by their whiteness and their fine cell structure. Polyurethane weather-stripping has a coarse spongy structure and it yellows very readily; even new polyurethane is rarely dead white. Polyolefins are chemically resistant to anything they are likely to encounter in the herbarium (except lubricating oils used on the hinges which might cause contact distortion) but they have poor recovery properties and gradually flatten under pressure. Because they are comparatively cheap and easily available, regular replacement is feasible.

The seals currently recommended by the CCI are dense, silicone rubber foams made without acetic acid. These are fairly expensive but appear to be durable. It is advisable to have the proposed silicone gasket checked because several of the ones recently tested at the ROM have not met the specifications.

Burning tests are a useful way to identify gasket materials, especially for distinguishing between cross-linked polyolefin foams and silicone foams which look and feel very similar. These tests should be carried out in the fumehood because the emissions from burning plastics can be extremely toxic. A very small piece of the plastic under test, held in insulated steel forceps, should be slowly brought to touch the edge of a flame for a few seconds before being pulled away. The odour from burning plastics has also been used as a means of identification but it is unreliable and unsafe. *Cross-linked polyolefin foams* usually begin to melt as they approach the flame, catch alight easily and melt, drip and bubble as they burn. On removal, the flame is self-sustaining and is often blue around the base. It generates a moderate amount of grey smoke. The ash is in the form of a fused bead. *Silicone foams* do not melt as they approach the flame. They ignite very readily producing a bright orange flame which is difficult to extinguish. The silicone glows as it burns. The residue is a fragile, grey, powdery ash in the shape of the original fragment. *Polyurethane* ignites easily, burning with a sooty, orange flame, lots of black smoke and burning drops of molten plastic. (Some urethanes have fire-retardants and are self-extinguishing on removal from the flame.) The residue is a black, powdery ash. *Neoprene* should be detectable with the Beilstein test. It does not melt as it approaches the flame. It does not ignite very readily, and it burns with black, sooty smoke. The flame is usually self-extinguishing. The residue is a black ash.

These are guidelines only as there will be variations caused by some additives. For example, gaskets filled with carbon glow orange when burned.

Herbarium Case Gasket Adhesives and their Residues

The problem with most gaskets is the adhesive, which often fails before the gasket material itself. Some companies have solved the problem by designing enamel cases with a fitted runnel to hold the seal, eliminating the need for an adhesive. Otherwise acrylic adhesives are usually recommended.

Removing old adhesives is extremely difficult; with age many of them have become resistant to all but the most toxic solvents. Heavy cases are not easily moved to an area of good ventilation where poisonous, flammable vapours can be controlled. Heating the old residue with a hot air blower to soften it, is likely to generate poisonous fumes.

In many cases the adhesive residues are not likely to be a future contamination problem — they have already done their worst! It is only necessary to remove residues which interfere with the proper fit and adhesion of replacement seals. Mechanical scraping to remove loose material may be all that is necessary. Old credit cards make excellent scrapers which do not damage enamel. Soft, oozing adhesives are sometimes easier to remove if they are stippled generously with baking soda (sodium bicarbonate) the day before.

CONCLUSION

In most cases it is not necessary, or possible, to replace all unsatisfactory plastic materials used for the storage of museum natural history specimens. Embrittlement or contamination of the specimen may have already occurred so changing to safer materials will not undo damage that has already been done; it can only reduce future damage. However, it is worth ensuring that all future purchases are carefully assessed to protect the collections and the budget. It is also advisable to check (immediately upon receipt) that inferior plastics have not been substituted for the items specified.

LITERATURE CITED

BLACKSHAW, S. M. AND V. D. DANIELS. 1978. Selecting safe materials for use in the display and storage of antiquities. Pp. 1-9 in *Preprints of the 5th Triennial Meeting of the International Council of Museums*. ICOM, Paris 23:2.

BLACKSHAW, S. M. AND V. D. DANIELS. 1979. The testing of materials for use in storage and display in museums. The Conservator 3:16-19.

BYERS, S. C. 1993. What is the matter with polystyrene? SPNHC Newsletter, February 1993, pp. 2-3.

COXON, H. C. 1993. Practical pitfalls in the identification of plastics. Pp. 395-406 in D. W. Grattan, editor. *Saving the Twentieth Century: The Conservation of Modern Materials*. Canadian Conservation Institute, Ottawa.

DAWSON, J. 1984. Effects of pesticides on museum materials: a preliminary report. Pp. 350-354 in S. Barry, B. R. Haughton, G. E. Llewellyn, C. E. O'Rear, editors. *Biodeterioration 6. Postprints of the Contributions to the International Biodeterioration Symposium*. C. A. B. International Mycological Inst., The Biodeterioration Society.

FENN, J. D. 1990. Guidelines for selecting display case materials. Museum Quarterly 18:3:23-30.

FENN, J. D. 1995. Secret sabotage: reassessing the role of museum plastics in display and storage. Pp. 38-41 in M. M. Wright and J. H. Townsend, editors. *Resins Ancient and Modern*. Scottish Society for Conservation and Restoration, Edinburgh.

HOPWOOD, W. R. 1979. Choosing materials for prolonged proximity to museum objects. Pp. 44-49 in *American Institute for Conservation Preprints of the 7th Annual Meeting*. Toronto.

ODDY, W. A. 1973. An unsuspected danger in display. Museum Journal 1:27-28.

ODDY, W. A. 1975. The corrosion of metals on display. Pp. 235-238 in *Conservation In Archaeology and the Applied Arts, Postprints of the Stockholm Congress*. International Institute of Conservation, Rome.

ROSSOL, M. 1995. Not all mothballs are created equal. Arts, Crafts, and Theater Safety 9(8):3.

SAKATA, Y. AND Y. YAMADA. 1979. Deterioration of clothes by using mothballs. 1. Influence of different kinds of mothballs used at the same time onto poly(vinyl chloride) fibres. Pp. 45-72 in *Yamaguchi Joshi Daigaku Kenkyu Hokoku Dai-2 bu* 4-5. [In Japanese]

TAYLOR, T. 1988. *Caveman Chemistry of Films*. Taylor Made Company, Lima, PA.

TETREAULT, J. 1993. La mésure de l'acidité des produits volatils. Journal of the International Institute for Conservation-Canadian Group 17:17-25.

WADDINGTON, J. AND J. FENN. 1988. Preventive conservation of amber: some preliminary investigations. Collection Forum 4:25-31.

WILLIAMS, S. 1988. The Diphenylamine Spot Test for Cellulose Nitrate in Museum Objects. Canadian Conservation Institute Notes, Ottawa 17/2.

WILLIAMS, S. 1993. The Beilstein Test: Screening Organic and Polymeric Materials for the Presence of Chlorine, with Examples of Products Tested. Canadian Conservation Institute Notes, Ottawa 17/1.

ZHANG, J., D. THICKETT AND L. GREEN. 1994. Two tests for the detection of volatile organic acids and formaldehyde. Journal of the American Institute of Conservation 33:47-53.

APPENDIX A

"Moth repellant" insecticides, like many other familiar household products, are much more dangerous than is generally realised:
• Camphor attacks the central nervous system (brain and spinal cord), the loss of smell is said to result from long-term, high level exposures.

- Dichlorvos (2,2-dichlorovinyl dimethyl phosphate) is a cholinesterase inhibitor (it interferes with nerve transmission) and among its chronic effects are depression of the immune system and memory loss.
- Naphthalene causes anaemia and attacks the kidney and the liver. Individuals with glucose-6-phosphate dehydrogenase deficiency and newborns are susceptible to haemolytic anaemia at low levels of exposure (e.g., in breast milk).
- Paradichlorobenzene causes respiratory and liver damage and the International Agency for Research on Cancer (IARC) has listed it as a category 2B carcinogen (possibly carcinogenic to humans).
- Thymol is a topical irritant, an allergen, and a cardiac depressant. Effects of long-term exposures are said to include liver and kidney damage.

Air quality standards were established as a guideline for healthy adults (Table 1). They do not apply to children, pregnant women, or individuals whose resistance may be impaired by other ailments or by medication. It has not yet been established whether combined biocide vapours aggravate the risk, but it is safer to err on the side of caution.

TABLE 1. Air quality standards set by the American Conference of Governmental Industrial Hygienists (ACGIH).

Biocide	TLV-TWA* ppm	TLV-STEL* ppm
Camphor	2	3
Dichlorvos	0.1	
Naphthalene	10	15
Paradichlorobenzene	10	
Thymol	Not established†	Not established†

*Threshold limit values (TLVs) based on time-weighted averages (TWA) over an eight hour day or short term exposure limits (STEL) averaged over 15 minutes, which should not be exceeded.
†This does not mean safe.

PART II

CONTEMPORARY ISSUES

FACING HERBARIA

CHAPTER 14

An Overview of Bar Code Applications and Issues in Systematics Collections

GEORGE F. RUSSELL

Abstract.—Bar codes have existed in commercial applications for decades. However, they have only found their way into systematics collections within the last ten years. It is still uncommon for natural history collections to use bar codes to help manage their specimens and specimen data. Bar codes were introduced to the United States National Herbarium of the Smithsonian Institution in 1985 and are now being used to track over six hundred thousand specimens. The future is even more exciting as international networks permit remote access to, and exchange of, specimen data. An overview of responses to a bar code survey is presented along with a discussion of future directions and standards of application.

INTRODUCTION

Bar code technology is becoming a popular tool for the management of systematics collections. Despite the fact that this technology was developed over fifty years ago, it has only begun to receive serious attention in the systematics community within the last decade. Although suitable for a variety of collections-related applications, bar codes are used primarily to provide an inventory and the means to track specimens for both large and small collections in museums, universities, botanical gardens, and culture collections.

This technology may not be the solution to every need in collections-based applications, but it is certainly a convenient and cost-effective answer to many of them.

Bar codes offer obvious advantages when used in systematics collections. They improve the accuracy and speed of access to a single specimen record, provide a single unique key to every specimen or specimen part in the collection, and offer a method of tracking specimens and providing accountability. They are also useful when integrated into software programs that collect and manipulate specimen records. The most effective way to integrate bar codes depends entirely on the nature and condition of the particular collection and the institutional requirements for managing specimens and data.

Most readers are familiar with bar codes. They are labels with black and white alternating bars found primarily on commercial products that, when "read" by an optical scanner, translate into a string of alphabetic or numeric characters. However, because in a store setting the result of this action is to separate you from your hard-earned cash, you may not have spent much time appreciating the wonder of bar codes or considering how and why they are used!

The overview of bar coding provided here is based on a combination of the results of an informal survey of bar code use in the systematics community conducted by the author, and ten years experience using bar codes at the United States National Herbarium of the Smithsonian Institution (US). Considering the speed with which institutions are now adopting this technology for use in their collections, there will certainly be many more new applications by the time this paper reaches print. The information provided is designed to orient those who are contemplating the use of bar codes in their collections. This overview will also raise issues that must be considered if we are to move toward recognized standards for bar coding in herbaria in particular and natural history collections in general.

THE BAR CODE

Bar codes have two essential aspects: the physical label and the coded information. The label consists of the substrate upon which the bars are printed, the ink with which the bars are printed, and the adhesive used to attach the label to the specimen. Substrates are mostly paper or cardboard, but sometimes plastic, aluminum, etc. Paper or plastic labels can be produced on sheets or rolls depending on need. Some paper labels are treated in various ways to provide greater longevity and some are equipped with a Mylar® coating to reduce wear and tear and improve "scanability." Whether labels are bought from a vendor or printed internally, it is critical that the quality of the label matches the requirements of your application. Some applications are short-term, whereas others require archival materials for long-term storage. As managers of systematics collections, we expect our specimens to survive for hundreds of years, so it is important that the quality of the materials be consistent with that goal. Herbarium specimens lend themselves to the use of this technology, providing a nice flat paper surface on which to apply a label. However, bar codes are just as easily applied to other surfaces such as microscope slides, glass jars, cardboard boxes, and photographic slides.

The quality of the adhesive is frequently overlooked when buying bar code labels. Adhesives are, however, the single material for which it is difficult to obtain specifications from bar code vendors. Except for a couple of applications reported in the survey that are temporary in nature, such as field labels and shipping labels, it is important that the adhesive be long lasting. The most commonly used adhesive appears to be an acrylic that has been implemented in countless library applications. I have not been able to obtain a life expectancy for it.

The coded information on the bar code has two components: the symbology and the coded value. The *symbology* determines the width, grouping, and numbers of alternating black and white bars on the label according to one of several industry standards. The symbology also dictates which character set is available for creating coded values. Common symbologies include Universal Product Code (UPC), used in many retail applications, and Code 39, which provides a complete set of numerics and upper case letters together with a small selection of other characters. The number of bars required to successfully represent a character depends on the symbology. There is, therefore, a direct relationship between the length of the coded value and the size of the label. In order to either squeeze more data onto a label, or attain a smaller label, the density of bars must be increased. The new, so-called three-dimensional Code 49 increases the density by allowing for multiple horizontal lines of vertical bars, and so packs more data into a smaller label space. Sophisticated scanners can deal with bars of various densities. The density of the bars is expressed either as characters per inch, or the width of the narrowest bar. Formerly, the equipment used to scan and interpret bar codes needed to be configured for a particular bar code symbology. Modern scanners are equipped with software that has overcome this limitation and are now capable of recognizing numerous symbologies.

The *coded value* is the actual value of a particular bar code as interpreted by the scanner and expressed using the character set permitted by the symbology. There are two strategies for assigning this value. It may either be a meaningful value such as the catalog number, specimen number, or some other pre-existing collection code; or it may be a number systematically generated by the computer. The latter has been referred to as a "dumb" value, a value meaningful only to the computer. Recent applications have developed a hybrid between these two, using a collection coden followed by a dumb value.

There are two optional additions to bar code labels: the *displayed value* and the *header*. The displayed value is simply the coded value interpreted into ASCII characters and printed on the label directly above or below the bars. This has two possible advantages. First, the displayed value can be used to enter the code manually if, for some reason, the scanner is malfunctioning or the label is flawed. The second advantage applies only to specimens that have not previously been accessioned or catalogued, or those for which the catalog or accession number has been selected for use as the coded value. In these cases, the displayed value can stand as the accession or catalog number and the specimen will not have to be separately stamped. For specimens which are already numbered, displaying a dumb value adds a second number to the specimen which has the potential to create confusion over which number to cite for publication purposes.

The second option is the *header*, a short text string which may be useful in identifying the herbarium or the particular collection in which the specimen is deposited. As with the displayed value, the header might also serve to confuse users by drawing attention away from the formal catalog number, if one exists. Thus, as new applications are developed, these two optional components should be critically assessed with regard to both the organizational requirements and the uses of the collection.

HARDWARE AND SOFTWARE

The wide variety of equipment that is available from commercial vendors for use with bar codes can be overwhelming to the uninitiated. When US first embarked on a program to integrate bar codes into collections operations, a contractor was hired to assist us. We learned two important things from this experience: that bar code technology is a good deal simpler than it appears and that our contractor made it appear more complicated. This was especially true of hardware. The two most common types of hardware configurations cited by respondents to the survey are described here.

A *wedge reader* is a unit that sits between the desktop computer and the keyboard. Attached to the wedge reader is a scanning device. The scanning device may be either a contact wand or a non-contact laser "gun". When a bar code is scanned, the wedge reader interprets the bar code label and sends the interpreted value to the keyboard port. The CPU accepts the data as if it were entered directly on the keyboard. This feature permits bar codes to be easily integrated into existing database applications. Wedge readers are fairly cost-effective devices available from numerous commercial sources. When using a wedge reader on low-end laptop computers lacking a keyboard port, the reader plugs into an available serial port and the data input is controlled by software which is designed expressly for laptops.

A *hand held bar code reader* acts as both a scanning device and a data processor. These units can contain object databases (partial or entire depending on their storage capacity) and programs to collect and manipulate data. They are especially useful when the application requires mobility but does not require the sophistication or capacity of a desktop computer. Hand held scanners should also provide a mechanism for porting data directly to and from desktop machines.

As discussed earlier, bar code labels may be purchased from commercial vendors or produced internally using a variety of possible *printers*. Inexpensive software is available to produce bar codes on both dot-matrix and laser printers. While this is a simple and cost-effective process, it may sacrifice some of the label and adhesive quality that are important to managers of systematics collections. More sophisticated and more costly printers are available to produce high quality bar code labels. However, only one of the respondents to the survey was printing in this manner. Most were either purchasing the labels or printing them using a laser printer.

THE US NATIONAL HERBARIUM EXPERIENCE

Bar codes were introduced to US in 1985 as a way of managing the hundreds of thousands of specimens for which we already had computer records. A bar code label was retroactively applied to each inventoried sheet, slide, or packet and the value of each bar code was painstakingly entered in each respective inventory record of the existing database. This retroactive bar coding was very labor intensive, but also very necessary. It had become clear that, as the number of specimen records in the database dramatically increased, our ability to maintain the integrity of the data was being compromised as the specimens were loaned, studied, and annotated.

When the bar code labels were first specified, both the displayed value and header options were employed, despite the concern that the header imparted greater importance to the label and might cause some users to cite the number in the literature. However, the last batch of labels purchased eliminated the header but retained the displayed value. Now when a bar-coded specimen is loaned, a flyer is included that distinguishes between the US catalog number and the bar code number and provides guidance to the user when citing the specimen. I think for US this was the best decision.

Managers of collections whose specimens have not been previously numbered may find themselves in a different situation; that is, whether to assume the value of the bar code as the catalog number. There is no apparent disadvantage to following this path. However, another scenario at US will demonstrate a potential problem. The higher plant specimens had been accessioned prior to barcoding and so were assigned different (dumb) bar code values at random. The cryptogamic specimens had not previously been numbered and so, once bar coded, assumed the bar code value as the catalog number. The conflict arises in those (admittedly rare) cases when visitors or borrowers use material from both collections. In the case of the higher plants, the borrower is cautioned to cite the proper catalog number which is stamped on the sheet, while in the case of cryptogams the borrower is allowed to cite the bar code numbers. Such inconsistencies should be avoided as much as possible.

The bar code label should be located on the sheet or other specimen so that it can be easily seen and scanned. Obviously, each type of object or specimen will dictate the most appropriate location. At US, the label is placed, whenever possible, at the bottom center of each herbarium sheet. The bottom right is generally reserved for the main specimen label and the bottom left is difficult to see or scan in folders that open on the right.

The selection of the coded value was another of our early decisions that could have benefitted from farsightedness. Because US was one of the first museum collections to employ this technology, there was little discussion of how our use of bar codes would affect outside users. We considered bar codes to be almost entirely an internal collection management tool. At that time computer networks had not begun to develop as they have today. Thus, a very simple eight-digit numeric value (using Code 39) was selected that allowed for both reasonable collection growth, that is, almost 10 million specimens, and a low density symbology that would still fit on a small label. Newly-developed bar code applications now frequently include a coden, abbreviation, or an acronym to designate the collection or institution in which the specimen is deposited.

OVERVIEW OF SURVEY RESPONSES

In 1994 and 1995, a few random requests were put out on network discussion groups for examples of bar code usage in systematics collections. The response was a bit less than anticipated. However, the diversity of the applications was impressive. A few observations follow, drawn from the 23 responses received, that might be useful to collection managers interested in implementing bar coding systems.

• Respondents included those with holdings of plants, fungi, minerals, vertebrates, geology cores, arthropods, gene bank vouchers, and anthropological objects.

- In some collections, the bar code represents a collective object such as a box of cores or a drawer of arrowheads, but mostly they identify an individual specimen. There is usually a one-to-one relationship between bar codes and computer records.

- The primary applications mentioned by respondents include transaction management (especially loans), inventory accountability, tracking distribution of duplicates (especially entomology and botany), processing field collections, and assisting with the production of specimen labels.

- Most respondents used Code 39, although there is an increasing use of Code 49, the so-called 3-D format, that allows for multiple horizontal lines of vertical bars.

- The vast majority of respondents began bar coding within the last two years.

- Most respondents configured their application using software and hardware recommended by consultants and vendors who guided them through the initial implementation.

THE FUTURE OF BAR CODES IN SYSTEMATICS COLLECTIONS

In 1988, I commented in a conference paper that, some day, another technology might well overtake bar coding as a means of managing herbarium specimens and their related automated data. While that may still happen, at the moment there is nothing even remotely comparable on a cost per unit basis. Passive Interrogative Transponder (PIT) tags or Supertags are just now beginning to receive some attention in the network discussion lists. While they hold tremendous promise for systematics collections, they are still too costly for most applications. However, the emergence of the Internet as a conduit for information has given these technologies a new dimension. A recent electronic discussion on the TAXACOM listserver centered on the possibility of assigning object values to systematics collections in such a way that every specimen in the world, or at least those with automated records, would be uniquely identified. Imagine being able to log onto a network application that allowed you to scan the bar code (or PIT tag) of a specimen that you borrowed from a collection halfway around the world to access information not available from the specimen label including: images of the collecting locality, light and scanning electron micrographs, or derived research data (chemical, physiological, anatomical, etc.). This is not a technical problem. Rather, the challenge is to develop a community strategy for creating a system of unique bar code values and agreeing upon a management authority to implement that strategy.

SYMPOSIUM

DESTRUCTIVE SAMPLING
AND MOLECULAR SYSTEMATICS:

ARE WE MOVING TOWARD A CONSENSUS?

Chapter 15

Guidelines for the Use of Herbarium Materials in Molecular Research

Emily W. Wood, Torsten Eriksson

and Michael J. Donoghue

Abstract.—Destructive sampling of herbarium specimens for morphological and anatomical studies has been with us for a long time, and most herbaria have implemented at least an informal policy to deal with such requests. With the advent of molecular systematics, sampling has taken on an added dimension and has forced herbaria to deal with some basic questions regarding this new use of specimens. Discussion of these issues among morphological and molecular systematists and the curatorial staff at the Harvard University Herbaria has resulted in the development of a general policy on sampling; in addition, it has raised questions regarding the storage of DNAs and associated molecular data that must be addressed by the botanical community as a whole. Although it is not yet possible to formulate a precise policy regarding the return and storage of DNAs, we hope that by providing a set of guidelines at this stage we will be better able to monitor activities in this area and work toward a community-wide agreement on an optimal solution.

From the earliest days of natural history collections, curators have dealt with the necessity of balancing conservation of those collections with maximum use for the advancement of science and society. Recognizing that these specimens provide a permanent record of the existence of individual organisms, curators and researchers using those collections have been mindful of maintaining the integrity of each specimen. In instances where removal or dissection of a portion of the material was essential, permission was usually granted by the curator on the merit of the individual request; such sampling has been an element of many plant morphological studies over the years. Within the last decade, sampling for morphological and anatomical work has been supplemented by sampling for molecular studies, where technical advances have made obtaining DNA samples from specimens not only feasible but practical. Here we discuss the ways in which the Harvard University Herbaria (A, AMES, ECON, FH, GH, and NEBC) have been dealing with the issues raised by this new use of herbarium specimens.

With the advent of molecular systematics, destructive sampling of herbarium specimens takes on an added dimension. Over the last decade, the development and refinement of laboratory techniques for the extraction of DNA have proceeded rapidly, enabling researchers to use smaller and smaller amounts of plant material in order to obtain sufficient DNA for analysis, especially polymerase chain reaction-based approaches (Loockerman and Jansen, 1996). It was not unexpected, therefore, that researchers began to turn increasingly to herbaria as a source of plant tissue; herbarium specimens are readily accessible, represent a wide variety of taxa, and essentially eliminate the necessity of collecting and identifying additional material.

Yet, while requests for destructive sampling for DNA studies are increasing, many herbaria have not given serious consideration to policies and procedures for handling such requests. Recent discussions of general collections policies by Cato (1993) and Cato and Williams (1993) underscore the need for institutions to address these issues before they become a problem. In the case of Harvard University Herbaria, the establishment of three new laboratories for molecular studies in the building, and the relatively active loans program already in place in the herbarium, were the prime incentives for establishing a policy statement to deal with such requests for removal of material from herbarium specimens.

The logical first step for us at Harvard Herbaria was to initiate a general discussion among morphological and molecular systematists and herbarium curatorial staff. These discussions raised several questions. How likely would it be for researchers to use herbarium specimens for molecular work? How much material was needed for a typical isolation procedure? Were herbaria running the risk of being perceived as simply warehouses for researchers interested in a ready supply of DNA? If sampling of specimens for DNA studies was allowed, could it be adequately monitored, and could the community insure that specimens of relatively rare taxa — which might be in high demand for broad-based phylogenetic studies — be protected? In short, the immediate concerns were whether DNA sampling was something that we wanted to support, and, if so, what policies should we adopt for dealing with such requests?

The first question was answered in our initial round of discussions. The consensus was that sampling for molecular studies was indeed an important new use of herbarium specimens, and that making those specimens available to the scientific community would represent a significant new contribution. At the same time, it was obvious that guidelines were necessary to insure that the collections were used properly. It was critical to establish how best to balance this single use — destructive by its very nature — with preservation of the material for future needs. Subsequent discussions focused on these issues, and our resulting policy statement (Appendix I) provides the basis for dealing with requests for destructive sampling of herbarium specimens, including molecular studies. Specifically, we require that researchers return to Harvard Herbaria some product of their work, be it SEM photographs of pollen, slides of wood, or a portion of the extracted DNA. For molecular studies, we require that a protocol be sent to us in advance, that the results be reported to us in writing, and, as with any loan, that the specimen be annotated with the identification accepted by the user as well as notation of what material was removed for study. Unless additional packeted plant tissue was collected specifically for a given kind of study (e.g., air-dried leaf material to be used expressly for DNA extractions), we normally only allow removal of material from a sheet once for a given category of study, such as pollen for SEM studies. In cases of large or complicated requests, we encourage researchers to seek their own funds to visit Harvard Herbaria and select specimens themselves, with removal of material and/or permission for removal given under the supervision and approval of the appropriate staff. Also, we encourage researchers to search available living collections first before any sampling of herbarium specimens is initiated. Several requests for material sent to Harvard Herbaria, especially for Asian taxa, have been at least partially filled with material from the living collections at the Arnold Arboretum of Harvard University. That resource is renewable, the material is usually ample, and our experience has been that arboreta and botanical gardens are quite willing to make their collections available.

STORAGE OF DNA SAMPLES

DNA samples and results are only now beginning to be sent back from researchers. A loan to the University of Arizona is illustrative: a loan of seven specimens of the tribe Galegeae (Fabaceae) was sent for study and sampling (Wojciechowski and Sanderson, University of Arizona, pers. comm.) following receipt of a formal loan request from the Herbarium, University of Arizona (ARIZ), a signed authorization form, and an extraction protocol from the researchers. The returned loan was accompanied by six DNA samples, precipitated and air-dried, and a detailed written summary of the methods and results, both positive and negative. The DNA samples are being stored at −80°C in our molecular laboratories, along with extractions from other plant groups (some dried and some in buffer) made by staff and students of the Herbaria.

The return of a portion of extracted DNA samples to Harvard Herbaria is perhaps the part of our policy that is most controversial and experimental. Recent discussions of this issue by Whitfield and Cameron (1994a, 1994b) and by Hafner (1994) underscore important considerations about the role of museums as repositories for specimens and their derived data. We agree with the recommendations of all the authors that molecular extracts from specimens be deposited in museums equipped with facilities for long-term storage. In the case of the Harvard Herbaria, the request for a portion of the extracted DNA to be returned to us is being made on a trial basis, with results to be evaluated periodically. Depending upon the long-term value of the samples, the frequency of requests for their use, and the efficacy of long-term storage, we may modify our policy to either drop the requirement altogether, make the parameters more specific, or consolidate samples to a more centralized storage institution. In summary, we view our current policy statement as a work-in-progress, subject to change as the parameters become better defined.

The fact that we are developing another collection — not unlike the herbarium itself, or the wood or slide collections — requires us to consider the following questions:

1) If we are in the business of storing DNA samples, what can we do to foster their use? Is it realistic to envision a coordinated effort among herbaria, or some subset of repository herbaria, in which DNA is stored, the information databased, and the DNA made accessible to other researchers on request?

2) Is it really possible to effectively store DNA extracted from herbarium specimens — or DNA extracted from live organisms — for long periods of time, in such a way that the integrity of a sample is maintained, and additional molecular data can be obtained?

With the return of DNA to Harvard Herbaria as a condition of our policy, combined with a significant level of in-house activity, we have begun to establish our own repository for DNA samples; other herbaria with associated molecular laboratories are in a similar position. A simple database, developed by the second author, links each DNA sample to the corresponding herbarium specimen (Fig. 1). Further developments of this database might include its availability over the Internet via Harvard Herbaria's World Wide Web site (http://www.herbaria.harvard.edu), and coordination with Harvard Herbaria's existing Type Specimen Database, which tracks specimens via a unique bar code number. However, the question remains whether we as individual institutions and herbaria should continue to pursue individual and yet parallel paths with respect to DNA storage. An alternative would be to designate some group of repositories within the botanical community which could serve as clearinghouses for sample requests (Adams, 1994; Hafner, 1994; Whitfield and Cameron, 1994a). The latter course seems preferable to us, and is also practical since many institutions may not have the resources to store DNA samples. In addition, access to information can be streamlined if there are a few central repositories; this would also make it easier to develop database standards and links.

FIGURE 1. Entry in DNA sample database.

SAMPLE

Sample nr 1
Loc_freezer | Donoghue Lab |
Loc_box 1
Sample type Raw DNA
Amount 100 uL
Extract. data Regular CTAB (2x) with PVP (1%)
Source | Herbarium specimen |
Responsible Torsten Eriksson
Received 94-05-10
GenBank nr U41381
Literature Eriksson, T., and M. J. Donoghue. 1995. Phylogenetic analyses of *Sambucus* and *Adoxa* (Adoxaceae) based on nuclear ribosomal ITS sequences and morphology. Amer. J. Bot. Suppl. 82: 128 (abstract 367).
Note

Taxon

Family Adoxaceae
Genus Sambucus
sp. australasica (Lindl.) Fritsch
ssp.
var.
form

Voucher specimen

Collectors Schodde
nr 5172
date 13 Dec. 1966
Herbarium A
Acc. nr.
Country Australia
Det. by

LINKS BETWEEN DATA AND SPECIMEN

In their discussion of the issue of return of DNA to the lending institution and subsequent storage, Whitfield and Cameron (1994a) cited "little coordination of the storage of molecular systematic data with more traditional taxonomic sources of information." This problem of data coordination is a serious one which has plagued conceptually linked yet physically isolated collections for years. Harvard's own wood collection is an example: in some instances, the link between the wood sample housed in the Bailey-Wetmore Laboratory and its corresponding voucher specimen in the herbarium has been lost. However, advances in computer technology and widespread use of the Internet and World Wide Web will allow links between molecular data and biological specimens to be developed so that the data can be made readily available to the community at large. Blake et al. (1994) summarized a recent meeting on the interoperability of biological databases, in which representatives of seventeen separate databases met to review technical and semantic issues pertinent to integrating data from various molecular sequence, citation, specimen, nomenclature, and phylogenetic relational databases, and to formulate mechanisms for queries among these databases. Although problems exist both from the technological (i.e., multi-database relational queries) and systematic (i.e., standardization of taxonomic names/classifications) perspectives, these efforts will advance the cause of ready access to biological data of all sorts. In the meantime, databasing efforts such as TreeBASE (http://phylogeny.harvard.edu/treebase/) for storage of phylogenetic trees and data matrices (Sanderson et al., 1994) and the Sequence, Sources, Taxa (SST) database (http://www.tigr.org/tdb/sst/sst.html) for linking gene sequences to voucher specimen data (Blake and Bult, 1996) can serve as repositories for information that is now accumulating at an astonishing rate. As Thomas (1994) noted, databases such as SST make a critical link between the derived molecular data and their underlying source, pointing "back to the final arbiter, a curated specimen in a collection."

LONG-TERM STORAGE OF DNA EXTRACTED FROM HERBARIUM SPECIMENS

Even if the botanical community were soon to reach a consensus on DNA repositories and database links, a fundamental question remains: have we still such a long way to go in perfecting the technical aspects of DNA extraction and storage that a community-wide effort is not yet appropriate? DNA from herbarium material is often degraded and in low concentrations, and it can degrade further over a short period of time (Jansen et al., this volume). Jansen has indicated that some DNAs from herbarium material will not amplify easily after freezer storage for one year, making re-isolation necessary. Some of these problems may be circumvented by the use of laboratory procedures yielding ultra-clean extracts, such as ultracentrifugation with cesium chloride. In any event, before centralized repositories are established, basic questions about the stability of current extractions must be answered. Laboratories involved in molecular extractions, including our own, need to initiate long-term experiments in which samples kept under a variety of controlled conditions are subsequently tested for usability. Until these issues are clearly resolved, our intention is to continue to solicit a portion of the extracted DNA from researchers as a condition of our loans; however, this policy will be reviewed often, and we may eventually concentrate instead on storage of *information* relating to a given specimen and its associated molecular data.

RESPONSIBILITIES OF THE MOLECULAR AND MUSEUM COMMUNITIES

To date, our experience at Harvard Herbaria with the practice of sampling for molecular studies has been a positive one. We have endorsed the use of herbarium specimens for such studies within a prescribed set of guidelines that were developed as mentioned above with the full cooperation of molecular and morphological systematists and the curatorial staff of the Herbaria. At the same time, we have strengthened our commitment to the concept that individual herbarium specimens are a non-renewable resource.

A variety of issues related to sampling still need to be addressed before we will be sure of the value of long-term storage of DNA samples. We have addressed the question of the physical integrity of DNA samples, but the problem of possible contamination owing to carelessness in the laboratory also needs more attention (Mueller, this volume).

All systematists have a responsibility to insure that voucher specimens exist for their work, and they must take every opportunity to renew the resource whenever they are in the field. Efforts to store silica-gel or air-dried material to supplement vouchered collections are currently underway at several herbaria today (Chase and Hills, 1991; Miller, this volume). Field collectors should make use of these techniques to collect leaf tissue as part of their regular collecting routine. Today's systematists, regardless of the nature of their research, need to continue to train their students in plant collecting procedures in order to insure well-prepared voucher specimens for studies of all sorts and adequate material for the future.

ACKNOWLEDGMENTS

Thanks are extended to the faculty and staff of the Harvard Herbaria who participated in discussions leading to the development of a policy statement; to the participants at the 1995 SPNHC workshop, "Managing the Modern Herbarium", who provided an excellent forum for discussion of issues surrounding sampling; to Bob Jansen, Robert Adams, and Michael Fay for valuable discussion and comments; and to Bil Alverson and John Beaman, who provided valuable comments on the manuscript. The second author's research was supported by grants from the Swedish Institute and the American-Scandinavian Foundation.

LITERATURE CITED

ADAMS, R. P. 1994. DNA Bank-Net: An Overview. In R. P. Adams et al., editors. *Conservation of Plant Genes II: Utilization of Ancient and Modern DNA*. Missouri Botanical Garden, 276 pp.

BLAKE, J. A., C. J. BULT, M. J. DONOGHUE, J. HUMPHRIES, AND C. FIELDS. 1994. Interoperability of biological data bases: a meeting report. Systematic Biology 43(4):585-589.

BLAKE, J. A. AND C. J. BULT. 1996. Biological databases in an electronic age: Access to and use of biological database. In J. D. Ferraris and S. R. Palumbi, editors. *Molecular Zoology: Advances, Strategies, Protocols*. Wiley-Liss, New York.

CATO, P. S. 1993. Institution-wide policies for sampling. Collection Forum 9(1):27-39.

CATO, P. S. AND S. L. WILLIAMS. 1993. Guidelines for developing policies for the management and care of natural history collections. Collection Forum 9(2):84-107.

CHASE, M. W. AND H. H. HILLS. 1991. Silica gel: An ideal material for field preservation of leaf samples for DNA studies. Taxon 40(2):215-220.

HAFNER, M. S. 1994. Reply: Molecular extracts from museum specimens can— and should—be saved. Molecular Phylogenetics and Evolution 3(3):270-271.

JANSEN, R. K., D. J. LOOCKERMAN, AND H.-G. KIM. 1999. DNA sampling from herbarium material: a current perspective. Pp. 277-286 in D. A. Metsger and S. C. Byers, editors. *Managing the Modern Herbarium*. Society for the Preservation of Natural History Collections, Washington, DC, xxii+384 pp.

LOOCKERMAN, D. J. AND R. K. JANSEN. 1996. The use of herbarium material for DNA studies. Pp. 205-220 in T. F. Stuessy and S. Sohmer, editors. *Sampling the Green World*. Columbia University Press, New York.

MILLER, J. S. 1999. Banking desiccated leaf material as a resource for molecular phylogenetics. Pp. 331-344 in D. A. Metsger and S. C. Byers, editors. *Managing the Modern Herbarium*. Society for the Preservation of Natural History Collections, Washington, DC, xxii+384 pp.

MUELLER, G. M. 1999. A new challenge for mycological herbaria: destructive sampling of specimens for molecular data. Pp. 287-300 in D. A. Metsger and S. C. Byers, editors. *Managing the Modern Herbarium*. Society for the Preservation of Natural History Collections, Washington, DC, xxii+384 pp.

SANDERSON, M. J., M. J. DONOGHUE, W. PIEL, AND T. ERIKSSON. 1994. TreeBASE: A prototype database of phylogenetic analyses and an interactive tool for browsing the phylogeny of life. Supplement to the American Journal of Botany 81(6):183.

THOMAS, R. H. 1994. Molecules, museums and vouchers. Trends in Ecology and Evolution 9(11):413-414.

WHITFIELD, J. B. AND S. A. CAMERON. 1994a. Museum policies concerning specimen loans for molecular systematic research. Molecular Phylogenetics and Evolution 3(3):268-270.

_____. 1994b. Authors' response to Hafner. Molecular Phylogenetics and Evolution 3(3):271-272.

APPENDIX 1

DESTRUCTIVE SAMPLING OF HERBARIUM SPECIMENS

HARVARD UNIVERSITY HERBARIA

Policy Statement and Authorization Form

This is to acknowledge your request for destructive sampling of herbarium specimens of Harvard University Herbaria. We ask that you read this policy statement and complete the agreement outlined below.

The collections of Harvard Herbaria are maintained with the goal of balancing preservation of the integrity of herbarium specimens with utilization for scientific research. While every effort will be made to accommodate researchers, decisions concerning destructive sampling of collections are made on a case-by-case basis.

As a rule, no material may be removed from specimens without prior consent of the Director or an appropriate member of the curatorial staff. Permission for removal of material is contingent upon adherence to the following guidelines:

1- Leaf material, pollen, spores, etc. may be removed from specimens *only* when there is adequate material available. Care must be taken not to damage the specimen.

2- Material may not be removed from type collections, or from taxa represented in the herbarium by less than 3 collections, except in rare instances, and then only by an appropriate staff member.

3- Each specimen must be annotated indicating the material removed, the nature of the study, the scientific name accepted by the researcher, the researcher's name and institutional affiliation, and the date. The Harvard University Herbaria should be cited in any resulting publication, a copy of which *should be sent here*.

4- In general, material may not be removed from an herbarium sheet for a second time, if the nature of the study is the same (i.e., pollen material for SEM, leaf material for DNA analysis, etc.). Exceptions will be made where additional material has been collected specifically for DNA studies.

5- Depending on the purpose of the material removed, the researcher must return to the Harvard Herbaria the following: a duplicate permanent pollen, spore, or leaf slide, an SEM photograph, a sample of the extracted DNA (see no. 6), etc. Such material will be housed in the Herbaria in a suitable place, cross-referenced to the specimen from which it was removed, and made accessible to other researchers upon request.

6- Requests for removal of material for DNA studies will be reviewed by the Curator of materials for molecular studies and approved by the Director. An extraction protocol must be submitted, along with an estimate of the amount of material needed. Results (both positive and negative) must be reported in writing; specimens must be annotated; GenBank or other applicable database accession numbers must be included or sent later; and a sample of the DNA must be sent to the Harvard University Herbaria. Samples obtained in this way will be properly stored and curated, and material may be provided to others for further study.

7- For large or complicated requests for material for DNA extractions or other studies, researchers will be encouraged to come to the Harvard University Herbaria, using their own funds, and select specimens for sampling themselves. Specimens will be selected and set aside by the researcher; removal of material will be made with supervision and approval of appropriate staff. Not only does this reduce the work required of the curatorial staff, but it allows the investigator to make more precise selections based on specimen age, material in packets, etc.

8- The Harvard University Herbaria have not maintained records on the history of specimen collection or treatment methods; materials are supplied with no warranty of any kind.

If you agree to accept the materials under the above conditions, please sign below and return this form to the Manager of the Systematics Collections, Harvard University Herbaria, 22 Divinity Avenue, Cambridge, MA 02138. For DNA studies, please enclose an extraction protocol. Upon receipt of confirmation, we will contact you concerning the dispatch of the material.

Accepted: _____
              Printed Name of Institution

Research Investigator: _____
                          Printed Name

              _____
              Signature

              _____
              Date

CHAPTER 16

DNA Sampling from Herbarium Material: A Current Perspective

ROBERT K. JANSEN, DENNIS J. LOOCKERMAN

AND HYI-GYUNG KIM

Abstract.—The use of herbarium specimens for molecular systematic investigations is rapidly increasing. Small scale isolation of DNA from as little as 25 mg of leaves using a variety of modifications of standard laboratory methods provides sufficient material for molecular systematic comparisons. The use of polymerase chain reaction followed by DNA sequencing are the most commonly used procedures for generating systematic data. Surveys of both molecular systematics labs and herbaria were undertaken to determine the extent to which herbarium collections are used and policies for sampling specimens. Current use of herbarium specimens accounts for less than 10% of the taxa examined but the majority of the labs anticipate increased use. Most labs are examining DNA sequence variation of several genes, especially the chloroplast encoded genes *rbcL*, *ndhF*, and *matK* and the internal transcribed spacer (ITS) regions of the nuclear ribosomal repeat. Although all labs prefer using DNA extracted from fresh material, herbarium specimens offer several advantages, including providing automatic vouchers and the possibility of examining inaccessible groups or those that are rare or extinct. The survey of herbaria indicates that curators are amenable to the use of collections but that clear policies must be developed for destructive sampling. These policies should include proper annotation of collections used (including GenBank accession numbers), avoidance of duplicate sampling of specimens, and acknowledgments of herbaria in any publications.

INTRODUCTION

During the past 15 years there has been a rapid surge in the use of DNA characters in systematics. Earlier studies were based entirely on the restriction enzyme approach (e.g., Palmer and Zamir, 1982; Doyle and Beachy, 1985; Sytsma and Schaal, 1985; Jansen and Palmer, 1988). Because these methods require large quantities of fresh plant material it was not feasible to use herbarium specimens. More recently, DNA sequencing has become the preferred source of molecular data because of the development of rapid, efficient, and cost effective methods. The development of the polymerase chain reaction (PCR; Mullis and Faloona, 1987) has had a profound impact on molecular systematics because it enables comparisons of DNA variation using minute quantities of plant tissue, including material from museum specimens and from fossils (Rogers and Bendich, 1985; Bruns et al., 1990; Golenberg et al., 1990; Soltis et al., 1992; Blackwell and Chapman, 1993; Sytsma et al., 1993; Thomas and Paabo, 1993; Taylor and Swann, 1994; Loockerman and Jansen, 1996).

There have been several recent reviews of the use of museum specimens for molecular systematic comparisons in a number of taxa (Rogers and Bendich, 1985; Bruns et al., 1990; Cano and Poinar, 1993; Thomas and Paabo, 1993; Taylor and Swann, 1994; Savolainen et al., 1995; Loockerman and Jansen, 1996). Most of these have focused on technical modifications for using preserved material for DNA studies. Only the most recent review by Loockerman and Jansen (1996) actually examined the extent to which herbarium material is used and the policies that have been developed for its use. Considerable changes in both the use and the policies for the use of herbarium material for systematic studies have taken place since it was written. Thus the intent of this discussion is to summarize and update the information reported by Loockerman and Jansen (1996).

DNA METHODS

There are a wide variety of DNA methods employed in molecular systematic studies which can be grouped into two classes:

structural changes and nucleotide substitutions. In plants, structural changes have been examined primarily in the chloroplast genome. They include inversions, gene losses or transfers, intron losses, insertions/deletions, and transpositions (reviewed in Downie and Palmer, 1992). Nucleotide substitutions can be examined indirectly by restriction enzymes or directly by DNA sequencing (reviewed in Palmer et al., 1988; Olmstead and Palmer, 1994). The basic methods for gathering these types of DNA characters from herbarium collections is the same as for fresh material. However, because less and lower quality DNA is obtained from herbarium specimens, certain methods cannot be used. For example, it is not possible to perform restriction site comparisons using the southern hybridization approach because one would generally not be able to recover sufficient quantities of high molecular weight DNA from herbarium specimens. Thus for restriction site analysis it is necessary to target certain regions of a genome and amplify those regions via PCR prior to performing restriction digests.

Preservation of Plant Material

Several studies have explored the effectiveness of different preservation methods for recovering intact, high molecular weight DNA from herbarium specimens (Doyle and Dickson, 1987; Pyle and Adams, 1989; Liston et al., 1990; Chase and Hills, 1991; Haines and Cooper, 1993). All of these studies concur that the method of preservation is more important than the age of the specimen. Material that has not been treated with chemicals or high heat, but rather, rapidly air-dried is most likely to yield usable DNA. Therefore chemical treatment is not recommended. Information regarding specimen preservation methods should be maintained to assist future researchers to decide if it will be feasible to recover usable DNA from a particular specimen.

DNA Extraction

Based on our own experience, DNA extractions are most successful when large-scale isolations are carried out on preferentially fresh material or large quantities of air-dried leaf material, that has been purified by ultracentrifugation in cesium chloride/ethidium bromide gradients. Thus we *always* try to obtain fresh leaves or large quantities of air-dried material before resorting to herbarium specimens.

Several papers have been published recently giving detailed methods for isolating DNA from herbarium specimens (Rogers and Bendich, 1985; Bruns et al., 1990; Taylor and Swann, 1994; Savolainen et al., 1995; Loockerman and Jansen, 1996). Most labs use a detergent-based DNA isolation procedure with hexadecyltri-methylammonium bromide (CTAB) being the most commonly employed detergent. Loockerman and Jansen (1996) provides details of the small-scale CTAB method used in our lab which had worked very well for most of the herbarium specimens we had examined. Subsequently, however, we have encountered difficulties in PCR amplification of DNA from a number of plant taxa. For some of these taxa, precipitation in high salt removes contaminating polysaccharides from the DNA and helps to alleviate the problem (Fang et al., 1992). The main problem when amplifying DNA from herbarium material is template purity. DNA can sometimes be amplified after purification by ethanol precipitation or GeneCleaning (Bio 101) with glass milk.

DNA isolated from herbarium specimens seems to degrade rapidly. The rate of degradation is probably no faster than it is for DNA isolated from fresh material. However, there is often very little high molecular weight DNA to begin with in isolations from herbarium specimens. Thus standard storage in TE (10 mm Tris [Trizma base], 1 mm EDTA [ethylenediaminetetraacetic acid]) buffer at 4° or −20° C is not sufficient to preserve the DNA for long periods of time. Often much of the DNA has degraded after one year, making it very difficult to successfully amplify it using PCR. If this observation is generally true, it clearly poses a challenge for herbaria planning long-term storage of DNA from herbarium specimens (Wood et al., this volume). Furthermore, some herbaria such as Harvard University have developed a policy that generally allows a specimen to be sampled for DNA only once. Since the DNA degrades quickly, this policy may not be practicable.

PCR of Herbarium Material DNA

Depending on the taxonomic group, we have generally been

successful with PCR amplification of DNA from herbarium material. Our most successful efforts have been in the tribe Tageteae, of the family Asteraceae, from which we have successfully amplified and sequenced DNA from herbarium material for 15 of the 17 genera following the procedures outlined in Loockerman and Jansen (1996). In contrast, our success rate for obtaining DNA from herbarium material from another tribe of the Asteraceae, the Mutisieae, after much difficulty, is below 10%. We obtain what appears to be usable DNA, however the impurities in the DNA do not permit its amplification. Purification of the DNA by the methods mentioned above, especially precipitation in high salt (Fang et al., 1992), sometimes facilitates successful amplification. Amplification of difficult DNAs by using Hotstart or nested PCR (Albert and Fenyo, 1990) has also proved successful.

USE OF HERBARIUM MATERIAL
BY PLANT MOLECULAR SYSTEMATISTS

A survey of 41 plant molecular systematics labs was carried out to determine the extent to which herbarium material was used for DNA studies (Table 12.1 in Loockerman and Jansen, 1996). Most of the labs (22 of 35 that responded) are using herbarium specimens, although in most cases the use is very low (less than 10% of the material examined). Investigators were generally using the modified CTAB method of Doyle and Doyle (1987) for DNA isolation and they were amplifying and sequencing a variety of chloroplast genes including *rbcL*, *ndhF*, *matK*, and the internal transcribed spacer (ITS) regions of the nuclear ribosomal repeat. Most labs were using angiosperm collections but fern and moss specimens were also being used. The age of the specimens ranged from 0 to 120 years old, with amplifications from specimens less than 20 years old being most successful. Specimens were borrowed from 22 different herbaria but most investigators used material from the herbarium at their own institution.

Overall the results of this survey clearly indicate that there is substantial use of herbarium material in plant molecular systematics. Most respondents indicated that they had just begun exploring the use of museum specimens and that they expected their use to increase in the near future. They showed a clear preference for using fresh material and resorted to herbarium specimens only when attempts to obtain fresh material had failed.

HERBARIUM POLICIES FOR DNA STUDIES

A survey was also conducted of 55 herbaria to determine if they had policies for the use of their collections for DNA studies (Loockerman and Jansen, 1996). Responses were received from 28 herbaria and the vast majority (25 of 28) indicated that they do allow removal of small amounts of material from specimens. Nearly all have some sort of destructive sampling policy in place. The demand for material for DNA studies has been very low or nonexistent but is increasing. Over half of the herbaria do not permit material to be removed from types.

CONCLUSIONS

It is clear that herbaria represent an important and underutilized resource for molecular systematic comparisons. Many labs are just beginning to rely on DNA isolated from herbarium specimens, especially when fresh material is either no longer available or not easily obtained. Molecular systematists still prefer DNA isolated from fresh material because it is far easier to work with. The low quantity and quality of DNA that can be extracted from preserved collections limits the kinds of approaches that can be used to those that involve PCR. Furthermore, the success rate of PCR amplification is substantially lower when using herbarium material. Thus it seems likely that herbarium material will be used mostly to supplement taxon sampling for taxa that are extinct, rare, or very difficult to obtain in the field or botanic gardens. A variety of modified isolation methods and PCR amplification strategies have been developed that seem to improve the success rate for amplification of the lower quality DNA obtained from herbarium specimens. Further improvements of these methods as well as the development of new techniques will likely increase the usefulness of herbarium material for future DNA studies.

Molecular systematics labs need to be responsible in their use of museum collections for DNA studies. Permission must be obtained prior to sampling the specimens and the herbaria must be acknowledged in any publications. The request for permission should provide a justification of why the collections need to be used. The specimens must also be annotated with information on the DNA isolation method used and GenBank accession numbers of any gene sequences generated from the material.

Herbaria clearly must develop more informed policies regarding the use of their specimens for DNA studies. Some herbaria have instituted a policy that prohibits sampling a specimen more than once. This seems to be an excessive restriction in many cases for three reasons. First, only a small amount of material is needed each time, often less than 25 mg (or less than 1 cm^2). This is less than the amount of material needed for morphological comparisons. Second, the best DNA isolation method is often taxon dependent, thus one may need to attempt multiple isolations before obtaining usable DNA. Finally, the prospect for successful long-term storage of DNA isolated from herbarium specimens does not seem very hopeful.

In addition, some herbaria have contemplated requiring researchers to return an aliquot of the isolated DNA with the specimen for long-term storage. Although this is an attractive concept we believe that it is not a good idea at this time. As mentioned above, DNA from herbarium material does not seem to be suited to long-term storage by the standard methods. Furthermore, many molecular systematists would be concerned about contamination of the DNA during storage. This is a particularly critical problem if the DNA is to be used for PCR amplification.

In conclusion, we believe that herbarium specimens will play an important role in molecular systematic studies in the future. It is, therefore, essential for molecular systematists to use these important museum collections responsibly and for herbarium curators to provide workable policies that allow access to the collections without jeopardizing their utility for other kinds of biological research.

Acknowledgments

We thank all of the current and past members of the Jansen lab who have contributed greatly to our success in using herbarium material for DNA studies. We also thank the herbaria and molecular systematics labs that responded to our surveys on herbarium policies and use of herbarium specimens. RKJ expresses gratitude to Deb Metsger for organizing the symposium on Managing the Modern Herbarium. Finally, we acknowledge NSF for two grants that supported portions of this research (DEB-9318279 to RKJ and DEB-9411536 to DJL and RKJ).

Literature Cited

ALBERT, J. AND E. M. FENYO. 1990. Simple, sensitive and specific detection of human immunodeficiency virus type 1 in clinical specimens by polymerase chain reaction with nested primers. J. Clin. Microbiol. 28:1560–1564.

BLACKWELL, M. AND R. L. CHAPMAN. 1993. Collection and storage of fungal and algal samples. Pp. 65–77 in E. A. Zimmer, T. J. White, R. L. Cann, and A. C. Wilson, editors. *Methods in Enzymology*. Vol. 224. Academic Press, San Diego.

BRUNS, T. D., R. FOGEL, AND J. W. TAYLOR. 1990. Amplification and sequencing of DNA from fungal herbarium specimens. Mycologia 82:175–184.

CANO, R. J. AND H. N. POINAR. 1993. Rapid isolation of DNA from fossil and museum specimens suitable for PCR. Biotechniques 15:432–435.

CHASE, M. W. AND H. H. HILLS. 1991. Silica gel: An ideal material for field preservation of leaf samples for DNA studies. Taxon 40:215–220.

DOWNIE, S. R. AND J. D. PALMER. 1992. Use of chloroplast DNA rearrangements in reconstructing plant phylogeny. Pp. 14–35 in P. Soltis, D. Soltis, and J. Doyle, editors. *Molecular Systematics of Plants*. Chapman and Hall, New York.

DOYLE, J. J. AND R. N. BEACHY. 1985. Ribosomal gene variation in soy bean (*Glycine*) and its relatives. Theor. Appl. Genet. 70:369–376.

DOYLE, J. J. AND E. E. DICKSON. 1987. Preservation of plant samples for DNA restriction endonuclease analysis. Taxon 36:715–722.

DOYLE, J. J. AND J. L. DOYLE. 1987. A rapid DNA isolation procedure for small quantities of fresh leaf tissue. Phytochem. Bull. 19:11–15.

FANG, G., S. HAMMAR, AND R. GRUMET. 1992. A quick and inexpensive method for removing polysaccharides from plant genomic DNA. Biotechniques 13:52–56.

GOLENBERG, E. M., D. E. GIANNASI, M. T. CLEGG, C. J. SMILEY, M. DURBIN, D. HENDERSON, AND G. ZURAWSKI. 1990. Chloroplast DNA sequence from a Miocene *Magnolia* species. Nature 344:656–658.

HAINES, J. H. AND C. R. COOPER, JR. 1993. DNA and mycological herbaria. Pp. 305–315 in D. R. Reynolds and J. W. Taylor, editors. *The Fungal Holomorph: Mitotic, Meiotic, and Pleomorphic Speciation in Fungal Systematics.* CAB International, Wallingford.

JANSEN, R. K. AND J. D. PALMER. 1988. Phylogenetic implications of chloroplast DNA restriction site variation in the Mutisieae (Asteraceae). Amer. J. Bot. 75:751–764.

LISTON, A., L. H. RIESEBERG, R. P. ADAMS, AND N. DO. 1990. A method for collecting dried plant specimens for DNA and isozyme analyses, and the results of a field test in Xinjiang, China. Ann. Missouri Bot. Gard. 77:859–863.

LOOCKERMAN, D. J. AND R. K. JANSEN. 1996. The use of herbarium material for DNA studies. Pp. 205-220 in T. F. Stuessy and S. Sohmer, editors. *Sampling the Green World.* Columbia University Press, New York.

MULLIS, K. B. AND F. FALOONA. 1987. Specific synthesis of DNA in vitro via a polymerase-catalyzed chain reaction. Meth. Enzym. 155:335–350.

OLMSTEAD, R. G. AND J. D. PALMER. 1994. Chloroplast DNA systematics: a review of methods and data analysis. Amer. J. Bot. 81:1205–1224.

PALMER, J. D. AND D. ZAMIR. 1982. Chloroplast DNA evolution and phylogenetic relationships in *Lycopersicon.* Proc. Natl. Acad. Sci. USA 79:5006–5010.

PALMER, J. D., R. K. JANSEN, H. J. MICHAELS, M. W. CHASE, AND J. R. MANHART. 1988. Chloroplast DNA variation and plant phylogeny. Ann. Missouri Bot. Gard. 75:1180–1206.

PYLE, M. M. AND R. P. ADAMS. 1989. *In situ* preservation of DNA in plant specimens. Taxon 38:576–581.

ROGERS, S. O. AND A. J. BENDICH. 1985. Extraction of DNA from milligram amounts of fresh, herbarium and mummified plant tissues. Plant Mol. Biol. 5:69-76.

SAVOLAINEN, V., P. O. CUENOUD, R. SPICHIGER, M. D. P. MARTINEZ, M. CREVECOEUR, AND J.-F. MANEN. 1995. The use of herbarium specimens in DNA phylogenetics: evaluation and improvement. Plant Syst. Evol. 197:87–98.

SOLTIS, P. S., D. E. SOLTIS, AND C. J. SMILEY. 1992. An *rbcL* sequence from a Miocene *Taxodium* (bald cypress). Proc. Natl. Acad. Sci. USA 89:449–451.

SYTSMA, K. J. AND B. A. SCHAAL. 1985. Phylogenetics of the *Lisianthius skinneri*

(Gentianaceae) complex in Panama utilizing DNA restriction fragment analysis. Evolution 39:594-608.

SYTSMA, K. J., T. J. GIVISH, J. F. SMITH, AND W. J. HAHN. 1993. Collection and storage of land plant samples for macromolecular comparisons. Pp. 23–37 in E. A. Zimmer, T. J. White, R. L. Cann, and A. C. Wilson, editors. *Methods in Enzymology*. Vol. 224. Academic Press, San Diego.

TAYLOR, J. W. AND E. C. SWANN. 1994. DNA from herbarium specimens. Pp. 166–181 in B. Herrmann and S. Hummel, editors. *Ancient DNA*. Springer-Verlag, New York.

THOMAS, W. K. AND S. PAABO. 1993. DNA sequences from old tissue remains. Pp. 407–419 in E. A. Zimmer, T. J. White, R. L. Cann, and A. C. Wilson, editors. *Methods in Enzymology*. Vol. 224. Academic Press, San Diego.

WOOD, E. W., T. ERIKSSON, AND M.J. DONOGHUE. 1999. Guidelines for the use of herbarium materials in molecular research. Pp. 265-276 in D. A. Metsger and S. C. Byers, editors. *Managing the Modern Herbarium*. Society for the Preservation of Natural History Collections, Washington, DC, xxii+384 pp.

CHAPTER 17

A New Challenge for Mycological Herbaria: Destructive Sampling of Specimens for Molecular Data

GREGORY M. MUELLER

Abstract.—Specimens housed in herbaria have value as sources of DNA as well as of morphological and ecological characters. Several factors unique to fungal collections, compared to vascular plant herbaria, need to be considered by both curatorial staff and potential users. These factors include: the small size and fragile nature of sporocarps of many fungal groups; the taxonomic diversity of fungal herbaria which typically house specimens representing at least two kingdoms and many classes; and, the destructive sampling of virtually every specimen borrowed, including types, for essential micromorphological studies. Informal surveys of curators and molecular systematists were made to assess current herbarium policy, extent of use of herbarium specimens for DNA studies, and potential conflicts and problems. The respondents were also asked to identify areas requiring more research. None of the respondents were opposed to expanding the role of herbaria as a source of and repository for materials for molecular studies. In fact, both groups were concerned with the optimal conditions for collecting and preserving voucher specimens to increase their usefulness as potential sources of DNA, and with herbaria storing DNA extracted from herbarium specimens. The key to expanding the use of fungal herbaria for this purpose is for curators to create a balance between the long-term preservation of valuable specimens and their accessibility for study. To achieve this balance, new users of these resources must agree to conform to standard herbarium etiquette.

INTRODUCTION

The advent of rapid methods to extract and amplify DNA revolutionized fungal systematics. Analyses of restriction fragment length polymorphisms (RFLPs) and sequence data have enabled mycologists to develop strongly-supported phylogenies — necessary components of evolutionary biological research including systematics (e.g., Bruns et al., 1991; Kohn, 1992). Because many fungi cannot be grown *in vitro*, sporocarps, either fresh or preserved herbarium specimens, are often the only source of DNA for systematic and population studies. Other fungal groups can be readily cultured, but many times the species in question are only infrequently encountered. In these cases, herbarium specimens, or material from well-maintained culture collections, are essential.

This paper examines the topic of destructive sampling of specimens for molecular studies from the viewpoint of the mycological herbarium. The focus is on those issues that make both curating and using material from mycological herbaria unique as compared to vascular plant herbaria. These factors need to be taken into account as herbaria develop policies on destructive sampling of fungal material for molecular data. They also need to be considered by systematists who wish to make use of herbarium resources. A brief discussion of some of the unique aspects of fungal herbaria is followed by a discussion of the results of two surveys sent respectively to curators at the major North American mycological herbaria and select molecular systematists who work with fungi. Finally, a brief summary of the state of knowledge of DNA extraction techniques and the effect of preservation techniques on the stability of DNA in fungal herbarium specimens is provided. Mycologists have been discussing these issues for 10 years (e.g., Mueller, 1988; Rogers et al., 1989; Bruns et al., 1990; Miller, 1990; Reynolds and Taylor, 1991; Haines and Cooper, 1993; Rogers, 1994; Taylor and Swann, 1994), but questions remain and consensus on policy is yet to be achieved.

UNIQUE ASPECTS OF FUNGAL HERBARIA

There are at least six factors that distinguish fungal collections from vascular plant collections. Destructive sampling for micromorphological studies has been a standard practice in mycologists' routine taxonomic protocols since the 1800s. While flowering plant systematists may sometimes need to dissect a flower, virtually all examinations of fungal specimens require dissection. Spores and other features visible only in sections or squash mounts under the microscope are the primary diagnostic and informative characters in fungal systematics. Even in large fleshy fungi such as mushrooms, basidiospore characters and other microscopic features are crucial for identification and for developing classifications. For many microfungi, microscopic examination is essential at all levels of interpretation and comparative study. Thus, for our mycological collections to be used, curators expect that specimens will undergo destructive sampling. The extent of this sampling is usually at the discretion of the user and the material removed is normally lost to the herbarium. Some herbaria request that sections and microscope slides made from their specimens be returned, but enforcement of this policy is uncommon.

Due to the relatively simple morphology of many fungi, it is often difficult to obtain a well-resolved, strongly-supported phylogeny based solely on morphology. Discerning between homology and convergence can be very difficult when there are few characters and states available. Thus, most of the recently published phylogenetic studies on fungi are based in large part on the analysis of molecular data. While much of the material used in these studies is either newly collected or obtained through culture collections, some previously preserved herbarium materials are employed. It is likely that herbaria will increasingly be called upon to provide material for such studies.

Sporocarps of many groups of fungi are very small, and the extraction of even the small amount of tissue necessary for DNA analyses may severely damage the specimen. Thus, the use of herbarium specimens for DNA studies exacerbates the potential damage to these specimens due to destructive sampling for micromorphological analyses.

Mycological herbaria house diverse specimens representing at least two kingdoms and many classes. Techniques for extraction and amplification of DNA that work well for some groups may not work for other taxa.

An herbarium specimen is often composed of pooled sporocarps, presumed to be of the same species, collected in a general area on a particular day. Because the vegetative mycelium is usually imbedded in the substrate from which sporocarps emerge, discerning genetically distinct individuals is difficult, and collectors make an educated guess about the extent of an individual based on the distribution of sporocarps. Also, it is not uncommon for collectors to pool sporocarps at various stages of development, putatively from the same species from one sampling site, to make a "good" herbarium specimen. Thus, it is possible that sporocarps of genetically distinct individuals, or even morphologically similar species, may sometimes be deposited as a single herbarium specimen. This has critical implications for using preserved herbarium specimens to address population biology and species level questions.

Many fungi are involved in mutualistic or parasitic relationships with plants and animals. For example, fungi such as lichens, mycorrhizae, fungal and foliar pathogens, and insect pathogens such as *Cordyceps* are accessioned and mounted in herbaria with their plant or animal associate. Such dual kingdom samples provide unique challenges for extracting DNA from the target fungus.

HERBARIUM POLICIES ON
DESTRUCTIVE SAMPLING OF FUNGAL SPECIMENS

An informal e-mail survey of curators at the major North American mycological herbaria was made in May 1995 to assess the current status of herbarium policy on destructive sampling (see Acknowledgments for list of respondents). The following questions were asked:

Does your institution have a policy regarding destructive sampling of fungal specimens for DNA analyses? If yes, describe. If not, is your institution considering one?

Only the New York Botanical Garden (NY) and the University of Minnesota (MIN) have a written policy. Their policies are not specific to sampling for DNA nor are they specific to their fungal herbaria. Rather, these institutions have merely added this type of sampling to their standard destructive sampling policy. The NY policy specifically states that material cannot be removed from type specimens. However, because one must dissect a small amount of material from fungal types for microscopic examination, this policy, of necessity, is flexible. One other herbarium indicated that it would follow and endorse a blanket policy for fungal specimens if one were developed. For the most part, the decision to allow destructive sampling is left up to the curator. Several of the respondents explicitly indicated that they had no problem with using type specimens, provided that there is ample material and that non-type specimens or cultures could not be substituted. This is the policy at the Field Museum. In short, none of the herbaria that responded viewed sampling for DNA data any differently from sampling for micromorphological characters.

What data do you require, or wish for, upon the return of specimens? Annotations? GenBank accession numbers? Herbarium acknowledgment in publication? Other?

All of the herbaria want their material annotated and hope that they will be cited in publications. However, there was no standard set of data that curators wished to have provided with returning specimens. The following minimum data seem appropriate:

• each herbarium specimen examined should be clearly annotated and, when possible, individual sporocarps be appropriately tagged or marked in some non-destructive manner;
• the GenBank number should be provided if the data are published, but if not yet published, the number should be provided to the herbarium as soon as it is known;
• information on the procedure used for extraction should be provided whether it was successful or not (inclusion of negative data is especially important if DNA extraction was not successful in order to alert curators and researchers that a particular procedure did not work); and,
• the specimen should be cited in ensuing publications, similar to expectations for morphological studies.

Approximately how many requests for material for this purpose have you had in the last three years?

Most of the herbaria have not experienced a rush of requests for loans from outside users; most report one to three requests per year. Several herbaria report relatively high use by staff at their institution. In these cases, many of the specimens were collected by the researcher and were deposited with the knowledge that they would be used in molecular systematic studies.

Do you have any general comments or ideas on the subject?

Several curators mentioned the need to begin thinking about storing DNA extracted from type specimens. Alternatively, they suggested developing special collections of whole or parts of fungi that would be preserved and stored under the "optimum" conditions for maintaining high quality DNA. These special collections could be mostly consumed. Neither the logistics of curating these materials, nor a consensus on what constitutes optimum conditions, appears to be well-developed at this time.

USE OF HERBARIUM MATERIAL
BY MOLECULAR FUNGAL SYSTEMATISTS

A similar set of questions was also sent to a number of fungal systematists who use molecular data. Responses were received from 8 of the 11 systematists surveyed. The respondents represent people working on a diversity of fungi from large mushrooms to powdery mildews and *Fusarium* (see Acknowledgments).

How often do you use herbarium specimens as a source for DNA?

All but two of the workers have used herbarium specimens in their research. Four of the respondents indicated that they routinely use herbarium specimens.

Do you have a good rate of success?

Success rates varied greatly. Scott Rogers (SUNY, Syracuse) and Tom Bruns (UC, Berkeley) reported very high rates of success with material less than 10 to 25 years old. While others reported lower success rates, they still had satisfactory results with material collected relatively recently. All respondents have had inconsistent results with older material, but some have had success with material over 100 years old. Several of the respondents have been able to germinate spores from well-preserved specimens and have subsequently used the resulting cultures for DNA extraction. Spore germination is very taxon specific. All of the researchers mentioned that they prepared and deposited specimens in their institution's herbarium to serve as voucher specimens and future sources of DNA.

How were the fungal herbarium specimens that worked preserved? Do any preservation techniques seem to work better than others?

Warm air drying (42°C), drying in silica gel, and lyophilizing (freeze-drying) all seem to work well for preserving DNA. Several respondents suggested that lyophilizing may be the best of these methods, but no one has really investigated this thoroughly. There have been reports that lyophilized material does not make usable herbarium specimens since micromorphological characters drastically deteriorate within several years (Harold Burdsall, USDA, Forest Products Lab, Madison, Wisconsin, pers. comm.). This important issue requires further study.

How much material do you use (weight or size of piece)?

All respondents report using very small quantities of material, ranging from a single perithecium to a cube 2 to 5 mm per side (data on minimum weight were not provided), about the same amount required for a good microscopic analysis. However, several people reported that if the extraction did not work the first time, the process required repetition and, in turn, additional material.

Have you encountered any difficulties borrowing material from herbaria? If so, describe?

None of the respondents reported problems with obtaining material on loan. All expressed the feeling of strong responsibility to use the material sparingly and to store it safely.

Do you have any general comments or ideas on the subject?

Comments and suggestions centered primarily on the need for more information on optimum preservation, storage, and extraction procedures and on the need to collect material specifically for DNA studies.

To summarize both surveys, no major problems in using herbaria as sources of DNA were identified and both groups of respondents were supportive of expanding the role of herbaria as a repository of molecular data. Neither curators nor users indicated any problem in the current system regulating the use of herbarium specimens for destructive sampling. This may be due to the current infrequency of requests made to herbaria by outside users for loans of material expressly for acquiring molecular data. As these requests become more frequent, it may be necessary to further codify herbarium policy on the subject, especially as related to sampling type specimens. Finally, curators and molecular systematists agreed on the need to develop optimal conditions both for preserving and storing herbarium specimens to maintain their DNA and for the storage of extracted DNA. Similarly, they agree on the need to develop special collections intended primarily for consumption. Such material would serve as good sources of DNA and would help alleviate pressure on historical voucher specimens.

DNA STABILITY: PRESERVATION AND EXTRACTION TECHNIQUES

Preservation techniques

There is a growing body of knowledge regarding the effect of various preservation and storage techniques on the DNA content and quality found in mushrooms and related fungi (Haines and Cooper, 1993; Taylor and Swann, 1994). In general, appropriate preservation techniques for fungi are the same as for flowering plants (e.g., Doyle and Dickson, 1987; Pyle and Adams, 1989; Chase and Hills, 1991).

Haines and Cooper (1993) looked at the effects of insect retardation treatments, specimen drying procedures, and storage of collections on the amount and quality of DNA in herbarium specimens. They used the common "grocery store" mushroom, *Agaricus bisporus*, and subjected specimens to various treatments.

Preservatives — Mercuric chloride, often mixed in a solution with turpentine, was widely used to prevent insect damage to fungal specimens until the 1920s. Its use was discontinued because of health hazards to humans. Mercuric chloride has now been shown to also negatively affect the preservation of DNA in treated specimens. Haines and Cooper (1993) painted mercuric chloride on fresh specimens before drying them. Only 20% of the DNA was recovered from specimens treated with mercuric chloride as compared to fresh material. They also placed fresh specimens in liquid preservatives including FAA (Bouin's solution: formalin, alcohol, acetic acid), phenol, and 70% EtOH. In all cases, the amount and quality of DNA recovered was significantly lower than air-dried specimens. The best results were obtained by drying the specimens at 42 to 50°C. They were able to recover nearly the same amount of DNA from warm air-dried specimens as from the fresh controls. Chase and Hills (1991) report extracting high quality DNA from plant material dried in silica gel. The effect on DNA preservation of other commonly used pesticide compounds such as naphthalene has not been examined.

Long-term storage — Haines and Cooper (1993) saw evidence of DNA degradation in specimens that had been preserved for as little as two years. The process of DNA degradation (i.e., desirable high molecular weight nucleic acids breaking into small fragments) was complete in most specimens by 33 years. However, they were able to find usable DNA in a few old specimens, up to 90 years, but this was very rare. Bruns et al. (1990) were successful in amplifying and sequencing preserved herbarium specimens over 40 years old. Kerry O'Donnell and Scott Rogers (USDA, Peoria, Illinois and SUNY, Syracuse, pers. comm.) both have had success amplifying and sequencing material over 100 years old. These historical specimens included macrofungi (e.g., Agaricales and Pezizales) and microfungi (e.g., *Fusarium* and *Trichoderma*).

There are two especially important messages to be learned from these studies. First, except in rare instances, historical specimens, including older type specimens, probably will not yield usable DNA, at least with current extraction and amplification techniques. And, second, we need to develop storage regimes that will prolong the usefulness of DNA in newly collected specimens.

Extraction and Amplification Techniques for Various Taxonomic Groups

There have been a number of papers published on the utility of various extraction and amplification techniques for preserved plant and fungal specimens (e.g., Rogers and Bendich, 1985; Rogers et al., 1989; Bruns et al., 1990; Rogers, 1994; Taylor and Swann, 1994; Jansen et al, this volume). For the most part, the same basic procedures work for both groups. Workers at UC, Berkeley and at SUNY, Syracuse have designed detailed protocols for fungi that give uniformly good results with relatively new material. However, many groups of fungi and organisms previously classified as fungi (e.g., slime molds and water molds now classified as protists) have not been included in these surveys, so the usefulness of these procedures for some taxonomic groups has not yet been determined. Thus, it would be appropriate for curators to ask researchers who are requesting the loan of some groups of microfungi to document that they have worked out the necessary procedures prior to sending them material.

MAKING AND CURATING SPECIAL COLLECTIONS
FOR DNA ANALYSES

Several respondents to the herbarium questionnaire mentioned the need to begin curating dedicated specimens, or parts of specimens, specifically as sources of DNA. A considerable portion of the discussion after the oral presentation of this paper at the 1995 Workshop "Managing the Modern Herbarium" in Toronto centered on this topic. Nearly everyone, curators and users alike, agreed that curating material for future DNA extraction is an appropriate activity for herbaria.

There are, however, unanswered questions concerning the optimum method of preserving herbarium specimens for future DNA extraction. At this time, however, it appears that warm air-drying at 42°C, followed by storage in a cool dry place away from sunlight is the optimum method for most fungi. Slight variation in these methods might be necessary to maintain long-term spore viability in some fungi. This happens to be the standard procedure for most herbaria, so little change in herbarium protocol is required. Some changes in collection management databases will have to be implemented if herbaria desire to track the fate of extracted DNA and polymerase chain reaction (PCR) products obtained from their specimens. Similarly, databases will have to be modified if herbaria institute a policy of accessioning material for the expressed purpose of destructive sampling since these specimens will be fully consumed. These are not insurmountable issues, simply ones that need to be addressed as herbaria adapt to serving new user communities.

A similar topic revolves around the idea that herbaria store and curate DNA extracts of type specimens or other valuable specimens. Many herbaria do not have the facilities to store and curate frozen extracted DNA, but air-dried or lyophilized DNA should not pose too great a problem. However, is it appropriate for herbaria to require users to return leftover extracted DNA along with the specimen? Will subsequent researchers be satisfied with using someone else's extracted DNA of critical taxa, or will they want to re-extract to be certain of the material that they include in their analyses? Cross-contamination in the laboratory and in the herbarium are serious possibilities. Herbaria will need to disseminate guidelines for appropriate handling of PCR products, DNAs, and specimens to reduce PCR contamination of materials. Minimally, the borrower should be expected to physically separate pre-PCR and post-PCR operations, equipment, and reagents. Specimens and DNAs to be returned to the herbarium should under no circumstances be exposed to PCR products. If DNAs are to be stored and loaned by herbaria, then the danger of PCR contamination may warrant dispensing individual aliquots to borrowers that would not be returned to the herbarium. These are general questions, ones that go beyond mycological herbaria and should be addressed by the natural history collections community at large. The reduction of possible contamination of specimens needs to be a goal of herbarium practices, regardless of the outcome of discussions on the advisability of having herbaria store extracted DNA.

CONCLUSION

Mycological herbaria differ significantly from vascular plant herbaria in several ways. These differences, however, do not impede the use of mycological specimens as sources of DNA. Based on informal surveys of curators and molecular systematists, the current system regulating the use of herbarium material is working well. However, there may be a need to further codify rules governing destructive sampling of fungal herbarium material for DNA if the number of requests for loans solely for this purpose increases. Points to remember when formulating these guidelines include:

1. Historical specimens should be used only when newly collected material or cultures are not available. This makes sense from both a curatorial and a user's viewpoint since it is easier to extract usable DNA from freshly collected sporocarps and cultures. Exceptions to this policy would need to be considered for studies requesting older specimens to screen for data such as historical distribution of mating types and historical distribution of molecular characters such as introns, insertions, and substitutions.

2. Type specimens must be preserved for use by future systematists; these specimens should only be used for molecular data

when there is ample material and when there is some confidence that usable DNA can be extracted. The curator also needs to use his/her judgment in deciding if non-type specimens or cultures could be substituted in the study for which type specimens are being requested. There are few studies that require the use of type specimens. Examples would be attempts at reconstructing species-level phylogenies when one or more taxa are represented only by the type specimen, and some kinds of nomenclatural studies.

3. Herbaria have the responsibility to preserve material in their care; curators need the flexibility to make case-by-case judgments on which specimens can be sampled for DNA. When possible, molecular systematists should be referred to culture collections from which they can receive research materials without grinding up non-replaceable herbarium specimens.

4. Users must fully annotate each specimen and cite material in their publications (see recommendations given above in section on the survey of herbaria).

5. Efforts should be made to collect material for DNA analyses, especially of critical or rare taxa, and from exotic locales. This material can be in the form of preserved herbarium specimens or cultures (fresh or freeze-dried). At present, herbarium specimens should be preserved and housed following the recommendations given in the previous section. Research to develop optimum preservation and long-term storage methods for all newly accessioned material is needed.

6. Curators should dissuade borrowers from trying to amplify highly conserved rDNAs from moldy specimens. Many contaminants on herbarium specimens are fungi. This poses special challenges for PCR-based fungal studies because the contaminant is often closely related to the target, making selective amplification difficult.

7. Mycologists need to participate in ongoing discussions on the appropriateness and utility of curating extracted DNA.

Great strides have been made in developing protocols to extract a new data set, DNA, from fungal specimens. Herbarium curators must deal with this new data set as they have done for macro- and micro-morphological data, balancing the long-term preservation of historical and irreplaceable specimens with their accessibility for current and future study. To achieve this balance, those using herbarium specimens for molecular studies must conform to standard herbarium etiquette. Specifically, they must provide safe storage of the material while on loan, treat it with care and respect, provide annotations before returning the material, and, finally, cite the specimens and acknowledge the herbaria in publications.

ACKNOWLEDGMENTS

Many thanks to the following people who took the time to respond to my questionnaire on herbarium policies. Curators: Joe Ammirati, University of Washington; Jim Ginns, DAOM, Ottawa; Richard Korf, Cornell University; John Krug, University of Toronto; David McLaughlin, University of Minnesota; Don Reynolds, Los Angeles County Museum; Amy Rossman, National Fungus Collections, Beltsville, Maryland; and, Barbara Thiers, New York Botanical Garden. Molecular systematists: Meredith Blackwell, Louisiana State University; Kerry O'Donnell, USDA, Peoria, Illinois; Scott Rogers, State University of New York, Syracuse; Greg Saenz and Tom Bruns, University of California, Berkeley; Monique Gardes, University of Colorado; François Lutzoni, Duke University; and Wendy Untereiner, University of Toronto. I thank Deb Metsger for inviting me to participate in, and for co-organizing, the informative symposium held after the SPNHC meetings in June 1995. I also wish to thank all of the participants in that symposium, speakers and audience, for their helpful comments following my presentation and during the general discussion. Finally, John Engel, John (Jack) Murphy, Betty Strack, and Qiuxin Wu (all of the Field Museum), David Hibbett (Harvard University) and an anonymous reviewer made valuable comments and suggestions for this paper.

LITERATURE CITED

BRUNS, T. D., R. FOGEL, AND J. W. TAYLOR. 1990. Amplification and sequencing of DNA from fungal herbarium specimens. Mycologia 82:175–184.

BRUNS, T. D., T. J. WHITE, AND J. W. TAYLOR. 1991. Fungal molecular systematics. Annual Review of Ecology and Systematics 22:525–564.

CHASE, M. W. AND H. H. HILLS. 1991. Silica gel: an ideal material for field preservation of leaf samples for DNA studies. Taxon 40:215–220.

DOYLE, J. J. AND E. E. DICKSON. 1987. Preservation of plant samples for DNA restriction endonuclease analysis. Taxon 36:715–722.

HAINES, J. H. AND C. R. COOPER, JR. 1993. DNA and mycological herbaria. Pp. 305–315 in D. R. Reynolds and J. W. Taylor, editors. *The Fungal Holomorph: Mitotic, Meiotic and Pleomorphic Speciation in Fungal Systematics.* CAB International, Wallingford, UK. 375 pp.

JANSEN, R. K., D. J. LOOKERMAN, AND H.-G. KIM. 1999. DNA sampling from herbarium materials in molecular research. Pp. 277-286 in D. A. Metsger and S. C. Byers, editors. *Managing the Modern Herbarium.* Society for the Preservation of Natural History Collections, Washington, DC, xxii+384 pp.

KOHN, L. M. 1992. Developing new characters for fungal systematics: An experimental approach for determining the rank of resolution. Mycologia 84:139–153.

MILLER, O. K., JR. 1990. Use of molecular techniques, the impact on herbarium specimens and the preservation of related voucher materials. Abstract. Mycological Society of America Newsletter 41:29–30.

MUELLER, G. M. 1988. Old techniques with new possibilities: importance of herbarium-based taxonomy today and tomorrow. McIlvainea 8(2):5–6.

PYLE, M. M. AND R. P. ADAMS. 1989. *In situ* preservation of DNA in plant specimens. Taxon 38:576–581.

REYNOLDS, D. R. AND J. W. TAYLOR. 1991. DNA specimens and the 'International code of botanical nomenclature.' Taxon 40: 311–315.

ROGERS, S. O. 1994. Phylogenetic and taxonomic information from herbarium and mummified DNA. Pp. 47–67 in R. P. Adams, J. S. Miller, E. M. Golenberg, and J. E. Adams, editors. *Conservation of Plant Genes II: Utilization of Ancient and Modern DNA.* Monographs in Systematic Botany from Missouri Botanical Garden, Vol. 48. Missouri Botanical Garden.

ROGERS, S. O. AND A. J. BENDICH. 1985. Extraction of DNA from milligram amounts of fresh, herbarium and mummified plant tissues. Plant Molecular Biology 5:69.

ROGERS, S. O., S. REHNER, C. BLEDSOE, G. J. MUELLER, AND J. F. AMMIRATI. 1989. Extraction of DNA from Basidiomycetes for ribosomal DNA hybridizations. Canadian Journal of Botany 67:1235–1243.

TAYLOR, J. W. AND E. C. SWANN. 1994. DNA from herbarium specimens. Pp. 166–181 in B. Herrmann and S. Hummel, editors. *Ancient DNA.* Springer-Verlag, New York.

CHAPTER 18

Destructive Sampling and Information Management in Molecular Systematic Research: An Entomological Perspective

JAMES B. WHITFIELD

Abstract.—The small size and large numbers of insect specimens handled by entomological collections present special vouchering and information management problems. There is, at the same time, an increasing trend toward integration of the results of molecular and more traditional morphology-based systematic studies. As a result, entomological museums, journals, and systematists need to develop consistent policies for the deposition of voucher specimens and for the management and databasing of the information derived from these specimens. A number of challenges in the designation of voucher specimens from molecular research are discussed, and several recommendations are made to meet these challenges. The advantages of specimen-based data management are discussed; additional recommendations are made to facilitate the integration of molecular systematic data into general specimen-based taxonomic databases.

INTRODUCTION

Molecular systematic research on insects, as on other groups of organisms, is relatively recent. Only within the last two decades have systematists applied molecular data to phylogenetic analyses, although population genetic studies using allozymes, and physiological and developmental investigations using protein and DNA sequences date further back. The appearance and refinement of polymerase chain reaction (PCR) technology (Mullis and Faloona, 1987; Kessing et al., 1989; White et al., 1989; Innis et al., 1990; Simon et al., 1991) have revolutionized the efficiency and affordability of obtaining DNA sequences from insects, including both live and preserved specimens. A number of introductory guides and protocols are now available for PCR and DNA sequencing of insects (Cameron et al., 1992; Brower and DeSalle, 1994; Hoy, 1994; Simon et al., 1994). Consequently, the use of preserved museum specimens for molecular studies is increasing and will likely continue to do so because they provide access to a wider range of taxa and localities than might normally be available to a researcher and may also reveal short-term evolutionary or distributional changes.

In spite of this, most systematic collections have not developed standard protocols for the use of their specimens in molecular research. The lack of such protocols has become a significant issue because, unlike traditional morphological sampling techniques, the process of converting tissues to DNA or protein extracts may result in partial or full destruction of the specimen. Thus there is a need for new policies governing the use of specimens from systematic collections in molecular systematic research.

This paper reviews a number of factors affecting the use of specimens for molecular research from an entomological perspective, including destructive sampling, preservation and storage techniques, and the long-term viability of DNA extracts. It then turns to the need to retain fragments and collection data of destroyed specimens, and to link them with the data resulting from molecular research. The recommendations offered should be applicable to other collections of animals, plants, and fungi.

THE ISSUE OF DESTRUCTIVE SAMPLING

The concept of destructive sampling is not new. Specimens from insect collections have long been dissected to reveal genitalia, mouthpart, and internal characters, or for making slide-mounts of wings or other body parts to facilitate their study. In a few cases, specimens preserved in fixatives are loaned for destructive comparative studies of internal anatomy. Usually, the specimen is returned with its dissected portions either contained in a small vial pinned with the specimen (Fig. 1) or mounted on a microscope slide. Museums have generally required preparator code numbers and allied notebooks to associate the specimen with the extracted information and body parts, although no uniform format has been developed to manage these data.

FIGURE 1. Pinned specimen of a carabid beetle, with preserved genitalic dissection associated with it using a genitalia vial.

Specimens used for DNA or protein analyses differ from most of the applications mentioned above: portions are not simply removed from the original specimens, they may be pulverized for extraction of the relevant molecules. Except for analysis of the extracted molecules, the removed portions of the specimens are no longer useful. The extracted molecules are also ultimately consumed in the course of the research.

In theory, one can amplify target DNA from a small fragment of a modern or fossil specimen, preserving the remainder for other uses. In reality, a number of factors can affect the DNA yield of a specimen, reducing it to the point that an insect specimen may be fully consumed. Some specimen-preservation methods can either lower DNA extraction yields (Post et al., 1993; Dillon et al., 1996; S. Cho, UC, Berkeley, pers. comm.) or directly or indirectly affect PCR efficiency. DNA is best preserved in freshly-caught or freshly-frozen (–80°C) specimens. Refrigeration enhances the preservation of tissues for DNA and morphological analyses alike (Masner, 1994), thus the next best source of DNA is specimens which were killed in the field in 95–100% ethanol and kept at 4°C. Dried, pinned specimens generally yield low quality, degraded DNA, although those killed in ethanol appear to yield more DNA (Dillon et al., 1996) than those that have simply died in killing bottles. Formalin-preservation destroys the DNA entirely and is therefore not recommended. The method of specimen preservation will become less important as molecular techniques are modified to become more sensitive, and with more taxon-specific oligonucleotides for PCR. Eventually increasingly smaller fragments of nearly all types of insect specimens may be useful for molecular research using PCR.

It may not be necessary in all cases for fully curated insect specimens to be used for molecular research. Most entomology collections have bulk stores of unsorted material from trap collections that is preserved in 70-95% ethanol. If these samples are transferred to 100% ethanol and refrigerated, they will be valuable storehouses for molecular research, at least as valuable as determined pinned collections. At present, however, this practice is uncommon.

Some molecular techniques, such as those used for allozyme or restriction site analysis, require fresh or frozen specimens rather than alcohol-preserved material. Currently only a few museums (Hafner, 1994) maintain collections of frozen specimens or tissues at ultra-cold temperatures. However, the procurement of such specially preserved insect specimens (frozen and/or ethanol-preserved) by museums will surely increase in the future, as their value becomes clear. They will likely be maintained in specialized facilities, separate from the principal, pinned voucher collections, similar to those described by Engstrom et al. (this volume) for mammals, and will be available on loan for molecular research.

When museum specimens loaned for molecular research must be destroyed, in whole or in part, the researcher should justify the use of the material for this purpose. Moreover, the researcher must demonstrate experience in the appropriate molecular techniques and methods of analysis. For example, researchers might be required to submit a brief research proposal, including pilot results if available. This would allow the museum to make informed decisions, weighing the scientific benefits of the research against the long-term cost to the collection and its future users.

Museums should be considering each of these issues as they begin to formulate policies that will take them into the twenty-first century, as molecular techniques becomes fully integrated into systematic research.

DNA EXTRACTS AS VOUCHERS

There has been recent debate over the proposal that systematic collections request the return of DNA extraction aliquots to serve as vouchers for material used for molecular work (Hafner, 1994; Thomas, 1994; Whitfield and Cameron, 1994a, 1994b). Several museums currently require the researcher to submit aliquots of DNA from specimens borrowed for molecular investigations (e.g., Harvard University Herbaria, The Natural History Museum, London). Several other institutions are prepared to receive aliquots of frozen DNA and tissue samples for long-term storage (Dessauer and Hafner, 1984; Dessauer et al., 1990; Hafner, 1994). Hillis and Moritz (1996) report on long-term tissue storage. Despite this, there is no general consensus among molecular systematists and museum curators supporting the mandatory requirement of returning DNA aliquots to the lending institution. This is not to imply that extracts should be discarded after use, but rather that the issue of permanent storage may have to be evaluated on a case-by-case basis, depending on the long-term viability of the extract and on the ability of the museum to properly curate it. In this era of dwindling museum budgets, it is unrealistic to expect many, let alone all, institutions to commit to long-term, high quality storage of frozen material.

In time, we will be better equipped to assess the long-term prospects of vouchering DNA. For the time being, when assessing the long-term usefulness of stored DNA extracts, museum curators must consider several factors. First, a variety of DNA extraction protocols has been used with insects (e.g., Simon et al., 1991; Cameron et al., 1992; Cameron, 1993). These differ widely with respect to purification procedures and the stability of the DNA extract because there is often a trade-off between ease and efficiency of extraction on the one hand and long-term stability on the other. Second, the long-term viability of DNA extracts under ultra-cold conditions is not yet fully known. However, preservation of ethanol-precipitated and/or dried DNA extracts may be superior to resuspended products.

Until then, it is reasonable to expect molecular systematists to save DNA extracts whenever possible, report the storage location of specimen extracts to the lending institution, and reference the location of relevant extracts in scientific publications. An ultimately more permanent and taxonomically useful form of vouchering may be preserving and documenting those parts of the museum specimens not sacrificed for molecular extraction.

REMNANTS OF SPECIMENS AS VOUCHERS

Most specimens of plants or vertebrates are large relative to the small sample of tissue required for DNA extraction. In contrast, insect specimens are small relative to the size of the tissue sample needed for DNA extraction.

Small insects are customarily ground *in toto*, though often wings and other appendages are removed before grinding. These

FIGURE 2. Mounted remnants from a braconid wasp specimen used in a molecular phylogenetic study. A second label is attached below the collection data label, outlining the use of the specimen in molecular research.

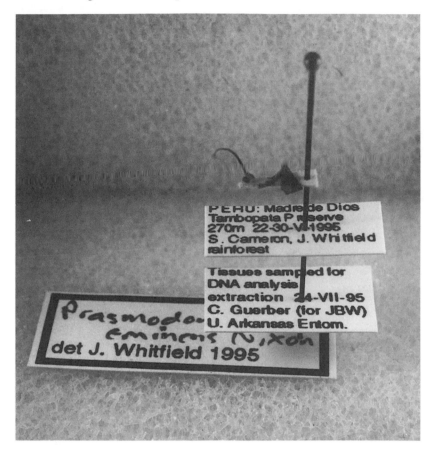

removed parts should be saved and retained with the original data labels (Figs. 2 and 3). However, small remains may not always suffice to verify the identification of a specimen at a later time. Therefore, it is best to save an additional specimen believed to be conspecific or from the same population, if available, which could in future be compared with the specimen parts.

With larger insects, a smaller portion of the specimen can be taken, such as the thorax or portions of it, or only one side of a bilaterally symmetrical animal may be used. The thorax contains a large amount of muscle tissue, rich in mitochondria and low in lipid

FIGURE 3. Small insect specimens are often ground for extraction in an Eppendorf tube using a small Teflon pestle. For good results, nearly an entire specimen the size of this braconid wasp may need to be used.

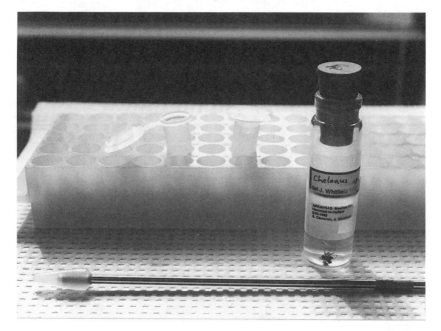

content, which produces relatively high yields of DNA. Furthermore, it contains fewer endosymbionts which can complicate phylogenetic analyses. Remnant portions of larger insects are probably sufficient as voucher specimens, as long as care is taken to preserve portions of the animal used in identification.

To date, many published molecular phylogenetic investigations have not documented remnant voucher material. Those authors that have used material borrowed from museums have generally been more conscientious. The fault is probably two fold. On the one hand, some molecular systematists are not familiar with vouchering protocols. On the other hand, museum collections have often not been adequately prepared to accept this material. To correct this, museums must develop the organizational framework for accepting remnant voucher specimens from all molecular systematic studies. The remnant voucher specimen, accompanied by its original label (collection locality and other information), should be incorporated into the main collection. This will allow systematists and ecologists to integrate information from the sampled specimen into a larger assessment of morphological and geographical relationships of an entire taxon, and to assess the need for further DNA sampling. An additional label (Fig. 4) should be kept on or with the DNA sample. It should include the essential information required to associate it with the remnant specimen voucher, including the researcher's name and address and the accession numbers for relevant DNA sequences stored in GenBank, EMBL or other databases.

FIGURE 4. Example of data labels associated with a voucher specimen from molecular phylogenetic research. The second label contains information related to its sampling, extraction, and resulting data. Note: the GenBank accession number is not yet assigned for this actual specimen; the number is hypothetical but of the form used by GenBank.

COSTA RICA: Guanacaste
Guanacaste Conservation Area
Pitilla Station, 500m el.
14-II-1995
J. B. Whitfield

Tissues sampled for
DNA analysis
extraction JBW 95-24
Univ. Arkansas Entomol.
GenBank Acc. UO8955

Diolcogaster
   xanthaspis - group

det J. B. Whitfield 1995

The scientific value of linking molecular data with relevant specimen vouchers stored in systematic collections that serve as permanent repositories cannot be overemphasized.

SPECIMEN-BASED DATA RETRIEVAL

Systematic collections, which already have data storage and retrieval systems in place, are a natural repository for vouchers (intact specimens or fragments thereof) from molecular systematic studies. Many collections are developing linked, computerized databases (Blake et al., 1994) and it would be valuable to also link data from molecular systematic research with taxonomic and phylogenetic databases (Thomas, 1994).

Many collections are moving toward the use of standardized bar code technology for tracking specimens and for linking specimens to other data. Bar codes greatly simplify the organization and retrieval of specimen-based data in computer databases. The entomological-museum community (through the Entomology Collections Network) has voted to use a standard, high density bar code, Code 49, for all entomological specimens. This is the first step to creating a single, compatible database system. The ultimate goal is to be able to electronically link all data of systematic interest to specific taxa.

Thus far I have presented the view that systematic collections should be fully involved in the storage and management of all information linked to specimens. Not all curators may agree with this outlook, either philosophically (Hafner, 1994) or for economic reasons, as tight budgets may limit a museum's role to that of archival storage. I argue that museums can and should manage specimen information and, likewise, play a major role in the development of taxon databases. In fact, the increasingly important role of museums in information management can be a strong justification for increases in funding.

CHANGES IN PUBLICATION REQUIREMENTS

Editors of journals and other publications can play an important role in specimen-based information management for molecular research. Many journals (e.g., *Proceedings of the National Academy of Sciences, Molecular Phylogenetics and Evolution*) require authors to submit sequence data to GenBank, EMBL, or an equivalent database and to report accession numbers for all taxa. All journals should make this a requirement.

Rarely are authors of molecular investigations required to give collection data for specimens, or state the location of voucher specimens. Yet such information is required of authors publishing taxonomic revisions based on comparative morphology, ecology, and behavior. This lack of attention to standard systematic practices may have become routine during the early days of molecular systematic research, when many analyses were concerned with questions of higher level relationships (e.g., kingdoms, phyla), or focused on well-known taxa (e.g., anthropoids). As it becomes increasingly common for molecular phylogenetic studies to address questions at lower taxonomic levels, and to include taxa of less certain status, we must apply the same standards for publishing as those that exist for more traditional taxonomic studies (i.e., collection-data documentation and voucher deposition). It is imperative that results of current molecular phylogenetic research be testable in the future. Hence, editors of journals that publish the results of molecular systematic research should require a high standard of specimen documentation.

SUMMARY OF RECOMMENDATIONS

Several guides have been developed to aid institutions in developing policies and guidelines for the use and management of collections (Cato, 1993; Cato and Williams, 1993; Hoagland, 1994). These guides deal broadly with the problem of destructive sampling. The following recommendations, together with those made in other chapters in this volume, offer additional guidance for the formulation of policies governing the use of specimens for molecular systematic research.

1. Each collection should develop criteria for allowing destructive sampling of their specimens.
2. Requesting researchers should be required to: a) submit a short proposal describing the project, presenting pilot results if available,

and estimating how much material will be needed for molecular extraction; b) submit a follow-up report on results and publications based on material, sequence databank accession numbers, and location of remaining extracts (if these are not also returned); and c) acknowledge the loaning and/or voucher institutions in any resulting publications.

3. The collection should have adequate storage and retrieval mechanisms for the above information.

4. Collections may require that an aliquot of each extract be returned either to them, or to another institution with facilities for long-term storage.

5. The standard voucher requirements applied to comparative morphological systematic research should also be applied to molecular systematic research. Voucher or remnant voucher specimens for each individual studied should be deposited in a permanent collection. This will facilitate future identification of and research on a particular taxon.

6. All journals publishing the results of molecular systematic research should require that each submission include complete collection data for specimens used, location of voucher specimens, and sequence database accession numbers (if relevant).

To ensure that the above recommendations can be achieved, communication between the systematic collections community and the molecular systematics community should be encouraged and enhanced.

ACKNOWLEDGMENTS

I thank Sydney Cameron, Chris Carlton, Paisley Cato, Soowon Cho, Rob DeSalle, Tim Dickinson, Mark Engstrom, Bob Jansen, Scott Lanyon, Deb Metsger, Jim Miller, Greg Mueller, Scott Shaw, Richard Thomas, Chris Thompson, and Emily Wood for sharing their views on current and prospective museum policies for molecular systematics. Soowon Cho, Lisa Vawter and especially Sydney Cameron offered many constructive comments on the manuscript, which greatly improved the result. The National Science Foundation (grant no. BSR-9111938), the US Department of Agriculture (9501893) and the Arkansas Science and Technology Authority (94-B-04) have funded my research in this area.

LITERATURE CITED

BLAKE, J. A., C. J. BULT, M. J. DONOGHUE, J. HUMPHRIES, AND C. FIELDS. 1994. Interoperability of biological data bases: a meeting report. Systematic Biology 43:585-589.

BROWER, A. V. Z. AND R. DeSALLE. 1994. Practical and theoretical considerations for choice of a DNA sequence region in insect molecular systematics, with a short review of published studies using nuclear gene regions. Annals of the Entomological Society of America 87:702-716.

CAMERON, S. A. 1993. Multiple origins of advanced eusociality in bees inferred from mitochondrial DNA sequences. Proceedings of the National Academy of Sciences of the USA 90:8687-8691.

CAMERON, S. A., J. N. DERR, A. D. AUSTIN, J. B. WOOLLEY, AND R. A. WHARTON. 1992. The application of nucleotide sequence data to phylogeny of the Hymenoptera: a review. Journal of Hymenoptera Research 1:63-79.

CATO, P. S. 1993. Institution-wide policy for sampling. Collection Forum 9:27-39.

CATO, P. S. AND S. L. WILLIAMS. 1993. Guidelines for developing policies for the management and care of natural history collections. Collection Forum 9:84-107.

DESSAUER, H. C., C. J. COLE , AND M. S. HAFNER. 1990. Collection and storage of tissues. Pp. 25-41 in D. M. Hillis and C. Moritz, editors. *Molecular Systematics*. Sinauer Associates, Sunderland, MA.

DESSAUER, H. C. AND M. S. HAFNER. 1984. *Collections of Frozen Tissues: Value, Management, Field and Laboratory Procedures, and Directory of Existing Collections*. Association of Systematics Collections, Lawrence, Kansas.

DILLON, N., A. D. AUSTIN, AND E. J. BARTOWSKY. 1996. Comparison of preservation techniques for DNA extraction from hymenopterous insects. Insect Molecular Biology 5:21-25.

ENGSTROM, M.D., R. W. MURPHY, AND O. HADDRATH. 1999. Sampling vertebrate collections for molecular research: practice and policies. Pp. 315-330 in D. A. Metsger and S. C. Byers, editors. *Managing the Modern Herbarium*. Society for the Preservation of Natural History Collections, Washington, DC, xxii+384 pp.

HAFNER, M. S. 1994. Reply: Molecular extracts from museum specimens can — and should — be saved. Molecular Phylogenetics and Evolution 3:270-271.

HILLIS, D. M. AND C. MORITZ, editors. 1996. *Molecular Systematics*. Second edition. Sinauer Associates, Sunderland, MA.

HOAGLAND, K. E., editor. 1994. *Guidelines for Institutional Policies and Planning in Natural History Collections.* Association of Systematics Collections, Washington, DC.

Hoy, M. A. 1994. *Insect Molecular Genetics: An Introduction to Principles and Applications.* Academic Press, San Diego.

Innis, M. A., D. H. Gelfand, J. J. Sninsky, and T. J. White. 1990. *PCR Protocols: A Guide to Methods and Applications.* Academic Press, New York.

Kessing, B., H. Croom, A. Martin, C. MacIntosh, W. O. McMillan, and S. Palumbi. 1989. *The Simple Fool's Guide to PCR. Version 1.0.* Manual distributed by the authors.

Masner, L. 1994. Effect of low temperature on preservation and quality of insect specimens stored in alcohol. Insect Collection News 9:14-15.

Mullis, K. B. and F. Faloona. 1987. Specific synthesis of DNA *in vitro* via a polymerase-catalyzed chain reaction. Methods in Entomology 155:335.

Post, R. J., P. K. Flook, and A. L. Milles. 1993. Methods for the preservation of insects for DNA studies. Biochemical Systematics and Ecology 21:85-92.

Simon, C., A. Franke, and A. Martin. 1991. The polymerase chain reaction: DNA extraction and amplification. Pp. 329-355 in G. M. Hewitt, editor. *Molecular Taxonomy.* NATO Advanced Studies Institute, H57. Springer Verlag, Berlin.

Simon, C., F. Frati, A. Beckenbach, B. Crespi, H. Liu, and P. Flook. 1994. Evolution, weighting, and phylogenetic utility of mitochondrial gene sequences and a compilation of conserved polymerase chain reaction primers. Annals of the Entomological Society of America 87:651-701.

Thomas, R. H. 1994. Molecules, museums and vouchers. Trends in Ecology and Evolution 9:413-414.

White, T. J., N. Arnheim, and H. A. Erlich. 1989. The polymerase chain reaction. Trends in Genetics 5:185-189.

Whitfield, J. B. and S. A. Cameron. 1994a. Museum policies concerning specimen loans for molecular systematic research. Molecular Phylogenetics and Evolution 3:268-270.

Whitfield, J. B. and S. A. Cameron. 1994b. Authors' response to Hafner. Molecular Phylogenetics and Evolution 3:271-272.

CHAPTER 19

Sampling Vertebrate Collections for Molecular Research: Practice and Policies

MARK D. ENGSTROM, ROBERT W. MURPHY

AND OLIVER HADDRATH

Abstract.—With the widespread use of protein electrophoresis in vertebrate systematics in the 1960s and 70s, several museums and universities established extensive collections of frozen tissues. Use of these special collections expanded exponentially with the development of the polymerase chain reaction (PCR) and improvements in obtaining nucleotide sequences. Moreover, using PCR, DNA sequences can now be routinely obtained directly from traditional voucher collections (skins, bone, alcohol-preserved specimens, etc.), accentuating the issue of destructive sampling of this material. Herein we briefly review: the development of special tissue collections for consumptive use; the rationale for and methods of collecting both voucher specimens and tissue samples amenable to genetic analyses; tissue collection storage and management; the suitability of voucher collections as direct sources of DNA; and policies on the consumptive sampling of special collections and destructive sampling of voucher collections.

Museums have the fundamental role of building and maintaining collections of biological specimens for documentation of biodiversity, studies of evolutionary pattern and process, and other evolutionary research. With the advent of the polymerase chain reaction (PCR; Mullis and Faloona, 1987) and other analytical tools for directly assaying genetic variation, use of these collections has rapidly shifted to include molecular-oriented research in addition to traditional morphological studies. In vertebrate collections, this change in orientation has been accompanied by the development of ancillary special collections used mainly for direct genetic assays. These special collections, such as frozen tissues, samples of isolated DNA, and ethanol-preserved tissues, have posed novel challenges for collection development and management, especially as they are ultimately designed for consumptive use. The rationale and intent behind these special collections contrasts sharply with those for traditional voucher collections. Accordingly, institutional policies for *consumptive* sampling of special collections will differ from the necessarily more restrictive policies associated with *destructive* sampling of voucher collections.

Extensive use of comparative molecular methods in population genetics and systematics of vertebrates has occurred only in the past 30 years (see reviews in Dessauer and Hafner, 1984; Honeycutt and Yates, 1994). In particular, the wide adaptation of protein electrophoresis in the 1960s and 70s (see Richardson et al., 1986; Murphy et al., 1996) provided the impetus for assembling frozen tissue collections. This was the first comparative technique that allowed routine, direct assessment of genetic variation within and between populations. Focus on the microevolutionary level, including the examination of large numbers of individuals to estimate parameters such as heterozygosity, polymorphism, and population subdivision led to the development of large, comprehensive tissue collections from wild populations (Baverstock and Moritz, 1996). More recently, these collections have proven invaluable for direct extraction of nucleic acids (DNA or RNA) used in a variety of studies, including molecular systematics and phylogenetics. For example, the number of comparative molecular papers using these collections and published in the Journal of Mammalogy increased from zero in 1954 to two in 1964, six in 1974, 14 in both 1984 and 1994, and 15 in 1996. Further, there has been a decided shift in techniques from karyology, immunology, and protein electrophoresis in the early part of this period to the current emphasis on comparative studies of DNA. These collections, together with the voucher specimens from which they were derived, also serve as baseline samples in forensic studies by law enforcement arms of agencies such as the US Fish and Wildlife Service, responsible for monitoring traffic in animals and their products (Dessauer and Goddard, 1984).

Concomitant with the acceptance, development, and routine application of molecular techniques, special collections have grown rapidly. In 1984, there were five North American collections with over 10,000 frozen tissue samples of vertebrates, all in the United States (Dessauer and Hafner, 1984). By 1994, there were 13 North American collections of this size and large collections were established both in Canada (Dessauer et al., 1996) and in México (F. Cervantes, Universidad Nacional Autónoma de México, pers. comm.). Several of these collections, such as the amphibian and reptile collections in the Royal Ontario Museum (ROM), now have more than 30,000 individual tissue samples (Dessauer et al., 1996). Growth of these collections has been facilitated by requirements of granting agencies. For example, the National Science Foundation and the National Geographic Society now routinely require that tissue samples be taken and deposited in major museums during the course of biological survey work.

More recently, the issue of destructive sampling of existing traditional collections of vertebrates as direct sources of DNA, where special collections are not available, has reopened an old problem for those responsible for maintaining the archival quality of those collections. Destructive sampling of voucher specimens for anatomical, physiological, medical, and other avenues of research has been a long-standing issue but has been brought to the fore by the sheer volume of requests (occasionally demands) from the molecular research community. All research collections are designed to be used, and curators and collection managers are faced with the conundrum of balancing present research needs with the charge of preserving the collections for future research (Cato, 1993). Although the prospect of using traditional specimens as sources of DNA has added an unforeseen research dimension, further justifying their archival value to sometimes sceptical administrations, this use must be carefully regulated. The widespread availability of special collections designed for such consumptive use in molecular systematics has helped to ameliorate, but not eliminate, these pressures for vertebrate collections.

The aim of this paper is to: 1) briefly review techniques for field collection, storage, and collection management of special collections for vertebrates (mainly frozen tissue collections); 2) discuss methodology and policies for destructive sampling of traditional, existing collections of voucher specimens; and 3) discuss institutional policies for consumptive use of special collections and destructive sampling of traditional voucher collections. It is hoped that this review will provide some background into the debate on consumptive sampling as it has evolved in the vertebrate museum community.

SPECIAL COLLECTIONS

Field Methods

Field preservation, transport, and storage of tissue collections were recently reviewed by Dessauer et al. (1996), and will be only briefly summarized here. For genetic analyses, tissues and their included polymers need to be collected and maintained in a biochemically active form. This necessity poses a series of challenges for field collecting, depending on location of the work and the types of molecular studies to be pursued. For vertebrates, samples are most often quick-frozen in either liquid nitrogen (LN_2: $-196°C$) or dry ice ($-60 °C$) in the field and later stored at ultra-low temperatures (near $-80°C$ or colder) upon return to the laboratory. For certain molecular approaches, such as studies of cellular DNA content (Sharbel et al., 1997) or isolation of whole, intact, mitochondrial DNA, initial freezing in LN_2 is required or highly preferable. Other approaches, such as DNA sequencing, pose fewer obstacles. Macromolecules can be preserved partially intact in a variety of solutions (see below), or even in dried pieces of skin or muscle (the latter enabling extraction of small segments of DNA from traditional specimens). To minimize financial and logistic constraints on field work and collection storage and maintenance, it is critical to determine at the outset which molecular approaches are to be supported by the collections. We routinely save tissue samples from most specimens collected.

A variety of tissue types can be collected in the field, and that variety may be affected by the intended molecular approach. For vertebrates, the array of tissues most commonly preserved includes blood, heart muscle, skeletal muscle, kidney, and liver, but may also include brain, spleen, stomach, eye (in fishes), and testes (in

birds and mammals). Tissues should be dissected from appropriately euthanized specimens as soon after death as possible. Skin
biopsies, blood and other tissues can sometimes be removed without killing the animal (e.g., Amos and Hoelzel, 1991; Seutin et al.,
1991; Whitmer and Barratt, 1996) but we recommend preparation of at least some specimens of each taxon as vouchers. For protein electrophoretic studies, the selection of tissues is critical
because the enzymatic expression of individual protein loci is often restricted to specific tissues. Tissue-specific expression of proteins varies widely among species (Matson, 1984; Murphy and
Crabtree, 1985; Murphy and Matson, 1986) with the result that
this information is sometimes useful in reconstructing phylogeny
(Fisher et al., 1980; Buth, 1984; Murphy and Crabtree, 1985). In
this regard, taking a greater diversity of tissues makes the collection more useful. For extraction of DNA, tissue type is less important, although mitochondrial-rich tissue such as liver or spleen is
best for extraction of whole mtDNA molecules. Longmire et al.
(1997) noted that for partially decomposed specimens of mammals, DNA in brain may be less subject to degradation than that
in some other tissue types.

Extended field trips in remote regions sometimes preclude the
use of LN_2 and other cryogenic options for tissue collection. Although freezing tissues remains the most reliable and versatile
method of preserving tissue for long-term storage, cryo-
preservation is not required for some molecular studies. DNA can
be preserved (although not wholly intact) in either 95% ethanol
or in 35% isopropyl alcohol. It is best to mince tissues to allow
quick penetration by alcohol, although whole specimens can also
be preserved in ethanol. Long-term stability of DNA molecules
may also be enhanced by adding ethylenediaminetetraacetic acid
(EDTA) which inhibits nucleases. Given that DNA is preserved in
alcohol, several institutions such as the Museum of Vertebrate Zoology, University of California, Berkeley, and the ROM ornithology collection, now maintain alcohol-fixed tissue collections.
Voucher collections never preserved in formaldehyde (e.g., Zoological Institute of St. Petersburg, Russia) also have been used as
sources of material for molecular research, although their value in
anatomical or histological studies may be compromised. Macer-
ated tissues initially preserved in lysis buffer in the field (Seutin et

al., 1991; Longmire et al., 1997), a preliminary step in DNA isolation, can be stored at room temperature, refrigerated, or frozen for several years prior to use. Finely minced tissues can also be saved in a saturated salt solution containing dimethyl sulfoxide (DMSO) and EDTA (Amos and Hoelzel, 1991; Seutin et al., 1991; Whitmer and Barratt, 1996) and stored at room temperature for six months or more. Preservation in either lysis buffer or DMSO salt solution produces larger yields of high molecular weight fragments of DNA than fixation in alcohol (Seutin et al., 1991).

Transportation and regulations for importation of animals and tissues to the United States are discussed in Dessauer and Hafner (1984) and Dessauer et al. (1996). In addition to obtaining any required research, collecting and/or export permits from the country of origin, Canadian regulations require a "Permit to Import Material of Animal or Microbial Origin Into Canada," obtained in advance from Agriculture and Agri-Food Canada (present fee, Can$21.00), to import tissues and specimens. Within Canada, written authorization is also required before exchanging imported tissues among institutions. If return airline schedules include a stop and customs clearance in the United States, the material must be declared there and all United States regulations satisfied (see Dessauer and Hafner, 1984) before proceeding.

Curation

Unlike traditional voucher specimens, it is usually impractical to re-number tissue vials once they have been returned from the field. Thus the field number initially written on the tissue container is the number used to identify the tissue and retrieve it from storage. Regardless of the numbering system, the sample should *always* be cross-referenced to the museum catalogue number of the voucher specimen from which it was taken.

The problem of field cataloguing has been handled in a variety of ways. Some institutions use a separate hard-bound field catalogue for tissues, wherein every tissue sample is assigned a unique sequential number, in addition to the field collector's number. These tissue catalogue pages are subsequently annotated with the permanent museum catalogue number of the voucher, thus cross-referencing the voucher with parts derived from it. A second system uses a single field number for both the tissue sample(s) and the voucher, and this number is recorded directly on the vial. On return to the lab, these samples are assigned another frozen tissue number and cross-referenced to the voucher collection. In both of these systems, three numbers are associated with each tissue sample: the collector's field number, the tissue catalogue number, and the catalogue number of the voucher specimen (Baker and Hafner, 1984).

Alternatively, the tissue catalogue can be eliminated altogether. In the ROM mammal and herpetology collections, collectors use pre-printed field catalogue pages and rolls of unique, sequential field numbers. These field numbers are used for both the voucher specimens and tissue samples. On return to the museum, the collection data are immediately entered into a temporary file and voucher specimens are assigned a permanent catalogue number. Thus, like the voucher specimen, only two numbers are associated with the tissue samples: the field number (used for retrieval purposes in the frozen tissue collection) and the permanent museum catalogue number of the voucher. Once field identifications are verified, the temporary database is updated and specimen information is sent to two permanent databases: the database for the main specimen collection and the frozen tissue database, each of which contains some unique fields (Woodward and Hylwka, 1993). The advantage of this system is that final determinations of the voucher and its collection catalogue number are automatically updated and cross-referenced to the tissue sample.

Frozen samples are permanently stored in cardboard boxes with dividers, which in turn are arranged in a system of metal racks in an ultra-cold freezer. Samples preserved in ethanol, lysis buffer or DMSO-salt solutions can be kept in the dark at room temperature, but DNA is more stable if tissues are kept cool. In the ROM ornithology collection, ethanol-preserved samples are stored at $-20°C$ for short intervals and between $-70°C$ and $-80°C$ for long term storage.

Two systems for sorting samples for retrieval are often used, either numeric, where samples are placed in order of field or tissue catalogue numbers (the system used in the ROM mammal collection), or taxonomic, where samples are arranged in the same order as the voucher collection (the system used by the ROM herpetology collection). The numeric system is more space efficient, as samples are simply added to the end of the numeric sequence. However, retrieval of specific taxa is more time consuming. Regardless of the system used, random gaps can occur in the boxes as samples are consumed.

Another challenge to the management of tissue collections is that samples are often completely consumed. To maintain a running inventory in the ROM mammal collection, we record the number of tissue vials originally present and update the database records any time a sample is granted to an outside investigator or used in-house. When the vial number equals zero, the sample has been totally consumed and it is effectively deaccessioned. However, records of how the sample was used and by whom are still maintained. One person (preferably a tissue collection manager) is assigned to maintain the tissue database, store and retrieve samples, and process inquiries and loan requests. Tissue grants are made only on the approval of the curator responsible for that collection.

DESTRUCTIVE SAMPLING
OF VOUCHER COLLECTIONS

Tissues do not necessarily need to be frozen or preserved in solution if collections are solely for studies of DNA. One of the great advantages of PCR is its ability to produce large quantities of DNA from very small numbers of target nucleic acids. Short fragments can be recovered from skin, bone, feathers, teeth, and other dried body parts that are hundreds and, in some cases, thousands of (or even more) years old (Pääbo et al., 1988; Ellegren, 1991; Hagelberg and Clegg, 1991).

Traditional voucher collections are now recognized by molecular systematists not only as the primary repository of information on morphological characters and relationships but also as storehouses of short strands of DNA. Although these specimens can be used, they present many problems for the recovery of DNA not experienced with frozen tissues. For example, the autolytic processes that follow cell death degrade nucleic acids. The resulting low concentration of target DNA is very susceptible to contamination from extraneous DNAs in the laboratory or from other sources. For example, Haddrath recently amplified a mitochondrial cytochrome b sequence from bird lice, only to find that the sequence recovered was actually that of the host (a kiwi) on which the lice had most recently fed. Inhibitors of the polymerase chain reaction that are found in traditionally preserved skins and tissues also tend to co-purify with DNA, often resulting in no or low yields (Hummel and Herrmann, 1994). Given these problems, frozen or otherwise preserved tissues from the special collections are much preferred as sources of DNA for PCR. However, for rare, endangered or, especially, extinct species, the original voucher specimens may be the best or only practical alternative. These factors should be considered in both the investigator's initial selection of source materials, and in the curatorial decision whether or not to grant a request to destructively sample specimens.

Although there are a variety of protocols for removing samples from museum specimens, several aspects are universal. When handling a specimen, latex gloves must be worn and any cutting should be done with sterile utensils. The smallest and least conspicuous tissue sample possible should be removed and then placed in a sterile plastic tube to lower the chance of contamination. It is best to remove two samples from separate places on the specimen. Both samples can then be processed independently to confirm the authenticity of any resulting DNA sequences and to minimize the possibility that any aliquots of DNA returned to the granting institution are contaminated. Double sampling and amplification is now required by some journal editors to confirm the repeatability of experiments involving ancient DNA. In the case of soft tissues, samples should be removed from subsurface areas when possible to limit the effects of contamination and action of preservatives (e.g., formaldehyde). In dried or mummified tissues, sampling near the extremities of the specimen enhances the chances of recovering DNA because these regions dehydrate quickly, limiting degrading autolytic processes.

The amount of material required from voucher specimens for PCR varies with the type of tissue being sampled. For soft tissues, such as dried skins, small pieces less than 0.1 g (1 or 2 mm^2) will often suffice (Thomas et al., 1990; Scott Woodward, Brigham Young University, pers. comm.), whereas for hard tissues, such as bone, between 0.5 g and 1.0 g usually is needed (Hagelberg and Clegg, 1991). One advantage in using bone is that longer fragments of DNA can be amplified from it, in some cases up to 1,000 base pairs (bp; Hagelberg et al., 1991). DNA in soft tissues is often more degraded and maximum fragment size is usually between 150 bp to 350 bp. In contrast, length of fragments in frozen tissues is limited only by the maximum size that can be amplified using PCR. State of preservation, rather than absolute age of the specimen, is the best indicator of the likelihood of successfully recovering a DNA sample.

We have been able to recover and amplify DNA fragments between 150–350 bp from skin samples in the voucher collection that are up to 100 years old, greater than 500 bp from bone up to 3,000 years old, and fragments greater than 350 bp from bone several thousand years old. For example, Haddrath used a 1 g sample of bone to isolate 400 bp fragments of DNA from a 10,000 year old New Zealand moa, and, using overlapping primers, reconstructed a 1,000 bp sequence from this extinct bird.

Methods of preservation which alter the chemical structure of vouchers may limit their utility in the recovery of DNA. For example, for large mammals it is very difficult to recover DNA from chemically-altered, tanned skins. In cases where large hides are to be tanned, we clip a piece of skin beforehand and store it along with the voucher skeleton. Likewise, dried museum skins are prepared without any preserving agents or insecticides, such as arsenic or mercuric chloride, and bones are rinsed only in water during their final cleaning. Although it is sometimes possible to recover DNA from specimens initially preserved in formalin, this is usually problematic. We have had little success amplifying DNA from well-fixed material, even when specimens have been subsequently transferred to ethanol for long-term storage.

Policies for Consumptive and Destructive Sampling

An important collections management issue is to determine under which circumstances it is appropriate to permit consumptive sampling from special collections or destructive sampling of traditional vouchers. Some balance must be struck between the need for access to collections by individual researchers and the long-term responsibility of the holding institution to maintain the future value and integrity of its collections. At the ROM, we are much more strict about destructive sampling of specimens in the main voucher collection than in loaning samples from special collections for consumptive use. For example, over a recent five year period, the ROM mammalogy collection has granted only three requests for destructive sampling of vouchers, whereas loans from the special frozen tissue collection comprised over 20% of all filled requests.

The issue of both consumptive and destructive sampling should be addressed in the form of a collections policy statement, such as that of the Museum of Vertebrate Zoology (MVZ), University of California, Berkeley (currently posted on the Internet at: http://www.mip.berkeley.edu/mvz/fcpolicy.html). A number of authors have discussed points that should be formalized in a written policy statement (e.g., Pääbo et al., 1992; Hafner, 1994; Whitfield and Cameron, 1994; American Society of Mammalogists, Systematic Collections Committee, F. Villablanca, in litt.), and Cato (1993) provided an example.

The following is a brief summary and amplification of points that we feel merit discussion in any formulation of a policy for both consumptive and destructive sampling.

1. Any request must be made in writing and should include a research proposal. Student proposals must be signed by the supervisor who must accept accountability for the material loaned.

2. The researcher should have demonstrated competency in the proposed methods, and positive results using a common, related taxon might be required before a request is granted.

3. Portions of samples that are not used should be returned to the collection at the conclusion of the project. Transfer of materials from the borrower to a third party should never be allowed, unless expressly authorized in writing by the lending institution. Such third party loans can lead to the loss of the connection between the voucher specimen and the tissue or subsequent DNA extraction. It is very important to avoid the loss of this link because it can result in the original lending institution losing both control of, and the rights to, information derived from its specimens.

4. Some institutions require that aliquots of DNA extractions be returned, to be stored by the lender (see discussion between Whitfield and Cameron, 1994; and Hafner, 1994). Return of

DNA extractions may prevent unnecessary resampling of the same specimens or tissue samples in the future since the extraction could be used instead. Our own experience with returned extractions is mixed, however, and we request this only for rare specimens, and then on a case-by-case basis. We have found that the long-term stability of DNA extracts can be affected by either the choice of reagents used to isolate the DNA or by the storage method. For example, while Chelex is very effective at extracting DNA from a wide range of cell types, the resultant high pH environment has been observed to affect the long-term stability of the DNA. Repeated freezing and thawing similarly compromises long-term stability. It should be noted that this can occur when specimens are stored in frost-free freezers due to the oscilation in temperatures of these units. There is also the possibility of the extractions being contaminated with extraneous DNA, a problem mostly found with extractions of ancient DNA. Unfortunately, these are often precisely the rare specimens which one would prefer not to resample.

5. For vertebrates, central depositories are being identified that are willing to accept and store returned extractions when the original lender does not have the facilities or personnel to do so (Hafner, 1994). It is unclear, however, how these central repositories will maintain the link between the original specimen and the extraction of PCR product, and under what circumstances these products might be loaned, given that the original lender should retain the rights to material derived from their specimens. Any repository system should be regulated by a disciplinary body (such as the Society for the Preservation of Natural History Collections or the Association of Systematics Collections) to insure that rights of the original collections are not inadvertently violated.

6. Collections differ in their policies regarding the scope of materials to be loaned (i.e., whether the lending institution is willing to serve as the major source of tissues and/or taxa for an outside study, or whether it is willing only to provide supplementary material to augment existing data). Some institutions, like the MVZ, will only provide supplementary material

whereas others, like some collections in the ROM, may provide the lion's share of the material required.

7. For destructive sampling of study specimens, some institutions allow known and qualified researchers to remove samples of skin, bone, feathers, etc., from specimens either under supervision in the collection at the lending institution or at the borrower's laboratory. Other institutions are more restrictive and insist on removing the samples themselves or in some cases, even performing the DNA extractions (e.g., zoological collections of the Natural History Museum, London). By choosing the specimen(s) and removing the tissue clips in house, the lending institution retains the most control over its material and insures that damage to specimens is minimized. This is the course most often taken at the ROM.

8. As noted by most authors of sequencing studies, DNA sequences should be submitted by the researcher to either GenBank or an equivalent database, and the accession numbers of these sequences should be returned to the lending institution to become a permanent part of the specimen record. Whether or not other results are required to be returned varies among institutions and types of studies. For example, in protein studies the individual genotypic frequency data might be deposited with the lending institution so that these data can be correlated with individual vouchers. This is particularly important if the original genotypic data are not published. However, accumulating, filing, and cross-referencing raw data has its limits. As pointed out by Hafner (1994), museums curate specimens, not reams of unpublished methods and results based on them. In the ROM we require, at minimum, the return of GenBank numbers and reprints of published papers, and ask for other data only in special circumstances. In all cases, any publications resulting from a study of borrowed material should both acknowledge the lending institution and list the individual specimens that were used.

9. Unlike loans of study specimens, reimbursement for shipping costs is usually required, particularly for frozen tissues which are sent by overnight courier. For large collections with relatively liberal grant policies, the cost of shipping soon becomes

insupportable unless the trade of material is evenly recipro-
cated. We occasionally waive this rule for starving graduate students. Some institutions also charge a retrieval fee, ranging
from $10 to $50 per tissue sample, to recover costs associated
with assembling and processing loans. Other institutions require that some tissue samples be received in exchange for any
tissues granted.

CONCLUSION

The distinction between *consumptive* sampling of special collections assembled for laboratory analyses and *destructive* sampling
of voucher collections is important and should be considered by
both molecular systematists and curators. Voucher collections are
preserved, processed, and stored to ensure their long-term value in
the documentation and study of biodiversity and related speci-
men-oriented systematic research, and their utility in molecular
systematics is a secondary consideration. Frozen and other special
tissue collections designed for consumptive use alleviate the need
to extensively sample traditional collections and remain the primary source of materials for molecular systematics. Institutional
policies for destructive sampling of vouchers are necessarily more
stringent than those regarding sampling of special collections designed for consumption. Although the advent of PCR has resulted
in important new uses for both traditional and special collections,
there is a continued requirement to balance the need for immediate use with the preservation of their archival value.

ACKNOWLEDGMENTS

This is contribution number 46 from the Centre for Biodiversity and
Conservation Biology, Royal Ontario Museum.

LITERATURE CITED

AMOS, W. AND A. R. HOELZEL. 1991. Long term preservation of whale skin for
    DNA analysis. Pp. 99–103 in A. R. Hoelzel, editor. *Genetic ecology of whales
    and dolphins.* Report of the International Whaling Commission, Special Issue 13.
BAKER, R. J. AND M. S. HAFNER. 1984. Curation of collections of frozen tissues.

Curatorial problems unique to frozen tissue collections. Pp. 35–40 in H. C. Dessauer and M. S. Hafner, editors. *Collections of Frozen Tissues: Value, Management, Field and Laboratory Procedures, and Directory of Existing Collections*. Association of Systematics Collections, Lawrence, Kansas.

BAVERSTOCK, P. R. AND C. MORITZ. 1996. Project design. Pp. 17–27 in D. M. Hillis, C. Moritz, and B. K. Mable, editors. *Molecular Systematics*. Second edition. Sinauer Associates, Inc., Sunderland, Mississippi.

BUTH, D. G. 1984. Applicability of electrophoretic data in systematic studies. Annual Review of Ecology and Systematics 15:501–522.

CATO, P. S. 1993. Institution-wide policy for sampling. Collection Forum 9:27–39.

DESSAUER, H. C. AND K. W. GODDARD. 1984. Value of frozen tissue collections for forensic studies. Pp. 12–13 in H. C. Dessauer and M. S. Hafner, editors. *Collections of Frozen Tissues: Value, Management, Field and Laboratory Procedures, and Directory of Existing Collections*. Association of Systematics Collections, Lawrence, Kansas.

DESSAUER, H. C. AND M. S. HAFNER. 1984. *Collections of Frozen Tissues: Value, Management, Field and Laboratory Procedures, and Directory of Existing Collections*. Association of Systematics Collections, Lawrence, Kansas.

DESSAUER, H. C., C. J. COLE, AND M. S. HAFNER. 1996. Collections and storage of tissues. Pp. 29–47 in D. M. Hillis, C. Moritz, and B. K. Mable, editors. *Molecular Systematics*. Second edition. Sinauer Associates, Sunderland, MA.

ELLEGREN, H. 1991. DNA typing of museum birds. Nature 354:113.

FISHER, S. E., J. B. SHAKLEE, S. D. FARRIS, AND G. S. WHITT. 1980. Evolution of five multilocus isozyme systems in the chordates. Genetica 52/53:73–85.

HAFNER, M. S. 1994. Reply: Molecular extracts from museum specimens can — and should — be saved. Molecular Phylogenetics and Evolution 3:270–271.

HAGELBERG, E. AND J. B. CLEGG. 1991. Isolation and characterization of DNA from archaeological bone. Proceedings Royal Society of London, B. 244:45–50.

HAGELBERG, E., L. S. BELL, T. ALLEN, A. BOYDE, S. J. JONES, AND J. B. CLEGG. 1991. Ancient bone DNA: techniques and applications. Philosophical Transactions Royal Society London, B. 333:399–407.

HONEYCUTT, R. L. AND T. L. YATES. 1994. Molecular systematics. Pp. 288–309 in E. C. Birney and J. R. Choate, editors. *Seventy-five Years of Mammalogy (1919–1994)*. Special Publication no. 11, American Society of Mammalogists.

HUMMEL, S. AND B. HERRMANN. 1994. General aspects of sample preparation. Pp. 59–68 in B. Herrmann and S. Hummel, editors. *Ancient DNA*. Springer-Verlag, New York.

LONGMIRE, J. L., M. MALTBIE, AND R. J. BAKER. 1997. The use of "lysis buffer" in

DNA isolation. Occasional Papers, The Museum, Texas Tech University 163:1–4.

MATSON, R. H. 1984. Applications of electrophoretic data in avian systematics. Auk 101:717–729.

MULLIS, K. B. AND F. FALOONA. 1987. Specific synthesis of DNA in vitro via a polymerase catalyzed chain reaction. Methods in Enzymology 155:335–350.

MURPHY, R. W. AND C. B. CRABTREE. 1985. Evolutionary aspects of isozyme patterns, number of loci, and tissue-specific gene expression in the prairie rattlesnake, *Crotalus viridis viridis.* Herpetologica 41:451–470.

MURPHY, R. W. AND R. H. MATSON. 1986. Gene expression in the tuatara, *Sphenodon punctatus.* New Zealand Journal of Zoology 13:573–581.

MURPHY, R. W., J. W. SITES, JR., D. G. BUTH, AND C. H. HAUFLER. 1996. Proteins I: Isozyme electrophoresis. Pp. 51–120 in D. M. Hillis, C. Morit, and B. K. Mable, editors. *Molecular Systematics.* Second edition. Sinauer Associates, Inc., Sunderland, Mississippi.

PÄÄBO, S., J. A. GIFFORD, AND A. C. WILSON. 1988. Mitochondrial DNA sequences from a 7000 year old brain. Nucleic Acids Research 16:9775–9787.

PÄÄBO, S., R. WAYNE, AND R. THOMAS. 1992. On the use of museum collections for molecular genetic studies. Ancient DNA Newsletter 1:4–5.

RICHARDSON, B. J., P. R. BAVERSTOCK, AND M. ADAMS. 1986. *Allozyme Electrophoresis.* Academic Press, Inc., Orlando, Florida.

SEUTIN, G., B. N. WHITE, AND P. T. BOAG. 1991. Preservation of avian blood and tissue samples for DNA analyses. Canadian Journal of Zoology 69:82–90.

SHARBEL, T. F., L. A. LOWCOCK, AND R. W. MURPHY. 1997. Flow cytometry: a precise and rapid method for genomic analyses of amphibian populations. In D. M. Green, editor. *Amphibians in Decline: Canadian Studies of a Global Problem.* Herpetological Conservation 1:78-86.

THOMAS, W. K., S. PÄÄBO, F. X. VILLABLANCA, AND A. C. WILSON. 1990. Spatial and temporal continuity of kangaroo rat populations shown by sequencing mitochondrial DNA from museum specimens. Journal of Molecular Evolution 31:101–112.

WHITFIELD, J. B. AND S. A. CAMERON. 1994. Museum policies concerning specimen loans for molecular systematic research. Molecular Phylogenetics and Evolution 3:268–270.

WHITMER, J. W. AND E. BARRATT. 1996. A non-lethal method of tissue sampling for genetic studies of chiropterans. Bat Research News 37:1–3.

WOODWARD, S. M. AND W. E. HYLWKA. 1993. A database for frozen tissues and karyotype slides. Collection Forum 9:76–83.

CHAPTER 20

Banking Desiccated Leaf Material as a Resource for Molecular Phylogenetics

JAMES S. MILLER

Abstract.—With increasing demand for sources of DNA for studies of molecular phylogenetics, the number of requests for destructive sampling of herbarium specimens is rising. To meet this new demand, botanical collectors can gather a small amount of additional leaf material specifically to support molecular studies with little added effort. A system for collecting, curating, and distributing desiccated leaf material as a source of DNA for studies of molecular phylogenetics is presented. The development of special collections of this material will alleviate pressure on herbarium voucher specimens and provide material of taxa that are poorly represented in herbaria.

INTRODUCTION AND BACKGROUND

Recent advances in molecular biology have allowed a re-examination of existing classification systems and the construction of robust phylogenies based on the analysis of nucleic acids. Sequencing of the chloroplast gene *rbcL* from a comprehensive sample of angiosperm taxa has resulted in many revised hypotheses of relationships among flowering plant families (Chase et al., 1993). These studies have since been expanded significantly by inclusion of a greater number of representative taxa and by extending study to include a wider variety of chloroplast, nuclear, and ribosomal genes.

Initial molecular systematic studies of the relationships of higher plant taxa often relied upon a few, readily-available representative species to characterize a plant family. Molecular systematic research has progressed to include more species and to focus on the taxonomic position of smaller, more poorly understood taxa. The relationships of these taxa often remain unclear, at least in part because species with restricted geographical distributions may be infrequently encountered by collectors, and so are often poorly represented in herbaria or other biological collections.

In an effort to maintain the integrity of rare specimens, many institutions have policies that prohibit removal of material from species represented by fewer than five specimens in their collections. Thus the taxa that are most desired, yet difficult to obtain, are those for which curators are least likely to approve destructive sampling. Providing an alternative means for ready access to DNA for these critical taxa will be essential to resolving our understanding of plant relationships in the future.

Plant material for molecular analysis may be field collected by the researcher or collaborating botanists, removed from existing herbarium specimens, or obtained from other assembled collections, such as living collections of botanical gardens, germplasm banks, and DNA banks. Studies of population structure or of the relationships between populations of a single species require intense field sampling to obtain large quantities of very specialized material. This task is often beyond the capacity of field botanists with other goals, however, the regional nature and specialized survey methods of these population studies make it both feasible and necessary for molecular biologists to undertake the collecting themselves.

In contrast, the study of relationships of higher taxa may require large numbers of species from diverse genera, families, and orders, often occurring in widespread geographical regions. It may be an impossible task for a single plant collector to amass the material required for these broad taxonomic studies. However, without access to material from a wide range of taxa, it will not be possible to resolve hypotheses of relationships of genera in large, widely distributed families or to understand the systematic position of uncommonly encountered genera or families.

The broad array of taxa required for such studies has usually been obtained by relying on field botanists to collect material for specific research projects. The success of relying on others to collect material for molecular studies depends on their willingness, the chance of their encountering the desired taxa, and their having both the necessary collecting supplies and the time to process the material. It often requires long periods of time to gather all of the taxa required for a study. Consequently, broad taxonomic studies that rely on obtaining field-collected material from many researchers often fail to secure all of the desired taxa.

Gaps in taxonomic datasets have recently been filled by removing material from herbarium specimens. The methods used to extract and amplify DNA from herbarium specimens of different ages and conditions have become more and more efficient over time (Jansen et al., this volume). Not only has the chance of successful extraction improved, but current methods require as little as 25 milligrams of leaf material to adequately supply the polymerase chain reaction (PCR) and DNA sequencing (Jansen et al., this volume). However, for the reasons indicated above, the specimens most likely to be requested in the future are often the very specimens for which curators are least likely to grant permission for removal of material.

One way to address the difficulty of obtaining appropriate plant material for destructive sampling is to bank plant material that has been collected and prepared specifically for molecular analysis. There are a number of advantages to a ready supply of material from a large variety of taxa that has been collected specifically for destructive sampling in support of molecular phylogenetics. Proactive gathering permits collection of sufficient plant material for multiple analyses, allowing repeated study of the same material by one or several researchers. This avoids the need to rely on DNA produced in other labs (many molecular biologists prefer to extract their own material) or to sample herbarium specimens more than once, a practice that is often discouraged or prohibited (Wood et al., this volume). Material gathered specifically for molecular analysis can be processed in a manner that will better preserve nucleic acids. Herbarium specimens, particularly when collected in the extreme wet conditions of parts of the tropics, are often processed and dried in ways that destroy their chemical integrity, such as drying at high temperatures or temporary preservation in alcohol or other solvents, resulting in poor quality DNA and/or low yields. Proactive gathering will also substantially increase the number of taxa that are readily available for use in research. Finally, such a special collection is not only of academic interest, but may also have commercial applications.

Before creating a special collection from which DNA could be extracted for either academic or commercial applications, prior informed consent from appropriate governmental regulatory agencies in each country must be obtained before any biological material can be collected and exported. A review of the legal and ethical issues is beyond the scope of this paper. However, the issues relevant to the transport of biological research material across international borders have been discussed elsewhere (e.g., Cragg et al., 1993, 1994a, 1994b; Laird, 1993; Posey, 1994).

COLLECTION AND CARE
OF SAMPLES FOR DNA ANALYSIS

Leaves of plants can easily be collected, preserved, and saved for later molecular analysis using relatively simple, inexpensive, and non-cumbersome methods (Rogers and Bendich, 1985; Doyle and Dickson, 1987; Pyle and Adams, 1989; Liston et al., 1990). Thus it is well within the normal capacity of botanists collecting herbarium specimens to gather leaf samples from at least a percentage of the taxa that they encounter and preserve them for molecular analysis.

It is important that each leaf sample taken for molecular analysis be vouchered by a high quality herbarium specimen to ensure proper identification of the plant material. For each sample, four to ten grams of leaf material are placed in a 10 x 15 cm reclosable plastic bag with at least 10 grams of desiccant for each gram of leaf tissue. Small leaves are collected whole, and larger leaves are torn into pieces that will easily fit into the bag; tearing apparently damages tissue less than cutting (Chase and Hills, 1991). A variety of

desiccants may be used; Adams et al. (1992) found silica gel and
drierite to be equally effective. Filling sample bags with desiccant
prior to field work eliminates the need to perform this task under
often uncomfortable conditions in the field and lessens the
chances of contamination. Botanists at the Missouri Botanical
Garden (MO) have found the less expensive silica gel produced
for the florist trade to be effective (Fig. 1). Indicator crystals,
which change color according to the relative moisture content of

FIGURE 1. Commercially available silica gel and re-closable plastic bags used for
collecting leaf samples. Preparing sample bags prior to fieldwork eliminates the
need for filling bags with desiccant under difficult field situations.

the desiccant, can be used both to ensure that the desiccant is dry
from the outset and to monitor the drying of leaves. If leaves do
not dry fully, the desiccant should be replaced with a fresh, dehy-
drated supply. The drying potential of exhausted desiccant can be
replenished by placing it for an hour or more in a 175°C oven, or
for longer periods over a plant specimen dryer.

Chase and Hills (1991) recommended removing most of the

silica gel for later reuse once leaves are fully dry. However, if the desiccant is not treated with temperatures high enough to destroy residual DNA, any small leaf fragments remaining could contaminate future samples, particularly with the use of PCR. Some packaged desiccants can be easily cleaned of potentially contaminating material, but they are much more expensive. In any case, once leaves have been fully dried, the majority of desiccant can be re-

FIGURE 2. Samples of desiccated leaves of Didymeles, the sole genus of the Madagascar endemic Family Didymelaceae, as they arrive at the Missouri Botanical Garden. The amount of material collected is adequate for numerous analyses and most of the desiccant has been removed to reduce weight during shipping.

moved to reduce the weight prior to shipping or to reduce the volume prior to long-term storage (Fig. 2). In the interval between collection and placement in long-term storage, efforts should be made to avoid exposing the samples to harsh environmental conditions which could denature the DNA, such as extended periods in full sunlight or high temperatures.

When samples are received at the institution where they will be housed, they should be organized numerically by collector (Fig. 3 and 4) to allow easy retrieval until they are fully labeled and are ready for incorporation into the permanent collection. Duplicate

FIGURE 3. DNA samples awaiting processing. Following arrival, each is checked for adequate desiccant and samples are kept in numerical order for each collector until labels have been prepared.

specimen labels should be prepared that cross-reference the voucher specimen to the sample of desiccated leaf material and this link should be clearly indicated in the specimen database. The choice of a 10 x 15 cm sample bag allows a standard specimen label to be placed inside the bag with the desiccated leaf material, and also permits storage in standard bryophyte cases (Fig. 5 and 6)

FIGURE 4. DNA samples stored temporarily in an upright freezer while labels are being prepared.

in a –20°C walk-in freezer. Storage in the original collection bag also reduces the risk of loss of material or collection data during processing. The quantity and condition of the desiccant in the bags should also be checked as labels are inserted and replaced if necessary.

FIGURE 5. Standard bryophyte specimen storage case placed in a walk-in freezer maintained at –20°C for long-term storage of DNA samples.

In the past, liquid nitrogen or ultra-cold freezers have been recommended for long-term storage of tissues (Dessauer and Menzies, 1984) and these recommendations probably remain true for extracted DNA. However, current practice suggests that a –20°C freezer is adequate to ensure that DNA is preserved in desiccated

FIGURE 6. Samples of desiccated leaf material in 10 x 15 cm re-closable plastic bags. Each easily accommodates a duplicate specimen label from the voucher insuring that complete data is available with the sample.

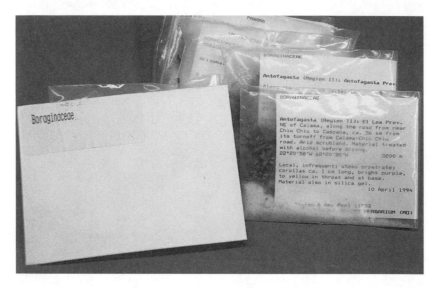

leaves (D. Nickrent, Southern Illinois University, pers. comm.). With eight drawers, two rows of samples per drawer, and an average sample capacity of 50 to 75 samples per row, each bryophyte case should hold an average of 1,000 desiccated leaf samples. At the Missouri Botanical Garden, desiccated leaf collections of bryophytes, vascular cryptogams, gymnosperms, monocots, and dicots are each filed separately. Samples within each of these groups are arranged alphabetically by family, and alphabetically by genus and species within each family. Families are separated by labeled rigid cards to facilitate access to the collections (Fig. 7). As requests are often for series of related taxa, this system allows easier retrieval than organizing material by individual collector.

When requests for material are filled, a photocopy of the specimen label from the packet is sent along with a sample of the desiccated leaves so that the researcher has complete specimen data. The presence of the duplicate label eliminates the need to also retrieve the associated herbarium voucher. It is important to track the distribution of each sample and make that information available so that individuals requesting material in the future will be

FIGURE 7. DNA samples in cold storage. Each of the two rows per drawer accommodates 50 to 75 labeled samples and families are separated by rigid cards for easy reference.

aware of other researchers who have used material from the same collection. Upon completion of sequencing studies, most institutions request that sequences be registered with GenBank. While some institutions are requesting that an aliquot of extracted DNA be returned for storage with the remaining sample material, the Missouri Botanical Garden is allowing researchers to retain all extracted and amplified DNA. However, the permission of the Garden is required for distribution of the DNA to a third party.

CONCLUSION

This system for collection and curation of DNA samples has the advantage that the leaves are collected into their permanent storage packets. Each sample bag can easily accommodate a specimen label so the full collection data remains with each sample. The size of these packets has been chosen to fit easily into a standard bryophyte storage case. Inclusion of a complete specimen label allows a photocopy to be distributed with each requested leaf sample without having to retrieve each individual voucher from the herbarium. Alphabetical storage of the samples allows for easy intercalation of new material and easy access to and retrieval of requested material. As requests are often for series of related taxa, this system allows easier retrieval than organizing material by individual collector.

Acknowledgments

The curatorial methodology described above was developed with input from many people but D. Brunner and J. Solomon played a major role in the development of this project at the Missouri Botanical Garden. A number of molecular biologists provided advice and I would particularly like to thank A. Colwell, W. Hahn, D. Nickrent, and K. Sytsma for their advice and encouragement. R. Magill, C. McMahon, and A. Tucker have assisted with computerization of the specimens and H. H. Schmidt provided the photos. F. Boncy, M. Merello, and H. H. Schmidt helped process the original samples.

Literature Cited

Adams, R. P., N. Do, and C. Ge-lin. 1992. Preservation of DNA in plant specimens from tropical species by desiccation. Pp. 135-152 in R. P. Adams and J. E. Adams, editors. *Conservation of Plant Genes. DNA Banking and in vitro Biotechnology.* Academic Press, San Diego.

Chase, M. W. and H. H. Hills. 1991. Silica gel: an ideal material for field preservation of leaf samples for DNA studies. Taxon 40:215-220.

Chase, M. W., D. E. Soltis, R. G. Olmstead, D. Morgan, D. H. Les, B. D. Mishler, M. R. Duvall, R. A. Price, H. G. Hills, Y.-L. Qiu, K. A. Kron, J. H. Rettig, E. Conti, J. D. Palmer, J. R. Manhart, K. J. Sytsma, H. J. Michaels, W. J. Kress, K. G. Karol, W. D. Clark, M. Hedren, B. S. Gaut, R. K. Jansen, K.-J. Kim, C. F. Wimpee, J. F. Smith, G. R. Furnier, S. H. Strauss, Q.-Y. Xiang, G. M. Plunkett, P. S. Soltis, S. M. Swensen, S. E. Williams, P. A. Gadek. K. J. Quinn, L. E. Eguiarte, E. Golenberg, G. H. Learn, S. W. Graham, S. C. H. Barrett, S. Dayanandan, and V. A. Albert. 1993. Phylogenetics of seed plants: an analysis of nucleotide sequences from the plastid gene rbcL. Ann. Missouri Bot. Gard. 80:528-580.

Cragg, G. M., M. R. Boyd, J. H. Cardellina II, M. R. Grever, S. A. Schepartz, K. M. Snader, and M. Suffness. 1993. Role of plants in the

National Cancer Institute drug discovery and development program. Pp. 80-95 in A. D. Kinghorn and M. F. Balandrin, editors. *Human Medicinal Agents from Plants.* American Chemical Society, Washington, DC.

CRAGG, G. M., M. R. BOYD, J. H. CARDELLINA II, D. J. NEWMAN, K. M. SNADER, AND T. G. McCLOUD. 1994a. Ethnobotany and drug discovery: the experience of the US National Cancer Institute. Pp. 178-196 in D. J. Chadwick and J. Marsh, editors. *Ethnobotany and the Search for New Drugs.* Ciba Foundation Symposium 185, John Wiley & Sons, Chichester.

CRAGG, G. M., M. R. BOYD, M. R. GREAVER, T. D. MAYS, D. J. NEWMAN, AND S. A. SCHEPARTZ. 1994b. Natural product drug development at the National Cancer Institute: policies for international collaboration and compensation. Pp. 221-232 in R. P. Adams, J. S. Miller, E. M. Golenberg, and J. E. Adams, editors. *Conservation of Plant Genes II: Utilization of Ancient and Modern DNA.* Monogr. Syst. Bot. Missouri Bot. Gard. 48.

DESSAUER, H. C. AND R. A. MENZIES. 1984. Stability of macromolecules during long term storage. Pp. 17-20 in H. C. Dessauer and M. S. Hafner, editors. *Collections of Frozen Tissues: Value, Management, Field and Laboratory Procedures, and Directory of Existing Collections.* Association of Systematics Collections, Lawrence, Kansas.

DOYLE, J. J. AND E. E. DICKSON. 1987. Preservation of plant samples for DNA restriction endonuclease analysis. Taxon 36:715-722.

JANSEN, R. K., D.J. LOOKERMAN, AND H.-G. KIM. 1999. DNA sampling from herbarium material: a current perspective. Pp. 277-286 in Metsger, D. A. and S. C. Byers, editors. *Managing the Modern Herbarium.* Society for the Preservation of Natural History Collections, Washington, DC, xxii+384 pp.

LAIRD, S. A. 1993. Contracts for biodiversity prospecting. Pp. 99-132 in W. V. Reid, A. Sittenfeld, S. A. Laird, D. H. Janzen, C. A. Meyer, M. A. Gollin, R. Gamez, and C. Juma, editors. *Biodiversity Prospecting: Using Genetic Resources for Sustainable Development.* World Resources Institute, Washington, DC.

LISTON, A., L. H. RIESEBERG, R. P. ADAMS, N. DO, AND Z. GE-LIN. 1990. A method for collecting dried plant specimens for DNA and isozyme analyses, and the results of a field test in Xinjiang, China. Ann. Missouri Bot. Gard. 77:859-863.

POSEY, D. A. 1994. International agreements and intellectual property right protection for indigenous peoples. Pp. 223-251 in T. Greaves, editor. *Intellectual Property Rights for Indigenous Peoples.* Society for Applied Anthropology, Oklahoma City.

PYLE, M. M. AND R. P. ADAMS. 1989. In situ preservation of DNA in plant speci-
    mens. Taxon 38:576-581.

ROGERS, S. O. AND A. J. BENDICH. 1985. Extraction of DNA from milligram
    amounts of fresh, herbarium, and mummified plant tissues. Pl. Molec. Biol.
    5:69-76.

WOOD, E. W., T. ERIKSSON, AND M. J. DONOGHUE. 1999. Guidelines for the use
    of herbarium materials in molecular research. Pp. 265-276 in Metsger, D.
    A. and S. C. Byers, editors. *Managing the Modern Herbarium*. Society for the
    Preservation of Natural History Collections, Washington, DC, xxii+384 pp.

CHAPTER 21

Recommendations on the Use of Herbarium and Other Museum Materials for Molecular Research: A Position Paper

DEBORAH A. METSGER

Abstract.—The symposium Destructive Sampling and Molecular Systematics: Are We Moving Toward a Consensus? included six papers each of which offered a different perspective on the use of specimens from museum collections for molecular systematic study. The symposium wrap-up session led to the construction of a position paper regarding the use of herbarium material in particular, and museum material in general, for molecular systematic research. This position paper acknowledges the role and value of museum voucher specimens as well as the opportunities afforded by molecular analysis. It recommends that the use of museum material for molecular systematic research be considered as a last resort rather than a first step. Where museum material *is* used, specific protocols are recommended for requesting material for sampling, and for acknowledging its use for molecular analysis. General principles are provided for authorization of sampling and for selecting specimens or other representative material for use. Suggested procedures are given for ensuring that voucher specimens are permanently linked to their molecular analyses. Herbaria and other museum collections are charged with documenting specimen care and usage to enhance the opportunity for future research use. Plant and animal systematists are charged with instructing all students in the use and value of systematics collections.

INTRODUCTION

The symposium Destructive Sampling and Molecular Systematics: Are We Moving Toward a Consensus? was held June 5–6, 1995 during the workshop "Managing the Modern Herbarium." The workshop was attended by 95 research scientists, curators, collections managers, and conservators representing 63 institutions from 6 countries. The symposium presented a range of disciplinary perspectives on the subject of destructive sampling of museum material for molecular systematic research. The concerns of herbaria and considerations for the development of policy were examined (Wood et al., this volume). Protocols for specimen usage and molecular techniques for four different types of organisms: vascular plants (Jansen et al., this volume), fungi (Mueller, this volume), insects (Whitfield, this volume), and vertebrates (Engstrom et al., this volume) were explored. The topic of proactive collecting of plant materials specifically for molecular systematic research was introduced (Miller, this volume). While proactive collecting to create special tissue collections has been conducted for several decades by zoological disciplines, it is a relatively new concept in herbaria. The wrap-up discussion focussed on the reciprocal responsibilities of herbaria/museum collections and molecular systematics labs to have policies and procedures in place that will safeguard the integrity of museum collections while ensuring their continued use in modern systematics research. The discussion culminated in the formulation of a position paper containing guidelines for the development of such policies and procedures. While these recommendations are directed primarily to herbaria and the botanical community, they apply to all museum systematics collections and their associated disciplines. We hope that relevant academic societies will endorse these recommendations, work toward the development of standards, and finally, encourage their members to adhere to them.

RECOMMENDATIONS FOR THE USE OF HERBARIUM AND OTHER MUSEUM MATERIAL IN MOLECULAR SYSTEMATICS

Participants in the workshop "Managing the Modern Herbarium" acknowledged the following basic principles regarding museum specimens and molecular systematic research:

1. museum collections provide permanent records of taxonomic concepts, biotic associations, habitats, and geographic distributions;
2. museum voucher specimens are the essential link between morphologically-based classifications and those based on macromolecular data;
3. recombinant DNA techniques increasingly provide opportunities to obtain phylogenetic data not only from new collections but also from museum collections;
4. this opportunity is bought at the price of destructive sampling of a part or all of specimens that may be unique examples of particular taxa, biotic associations, habitats, and geographic distributions.

With these principles in mind, the participants endorsed the following recommendations for the use of herbarium or other museum specimens in molecular systematic research:

The use of herbarium/museum material for the extraction of DNA or other compounds must be the last resort rather than the first step of a phylogenetic analysis. To ensure this:

- all requests for destructive sampling must be made through the curator responsible for the collection in question.
- it is the responsibility of the researcher to provide written justification for the proposed destructive sampling that makes clear that alternative sources, including culture or live collections, have been exhausted, and that the data to be obtained by destructive sampling warrant the sacrifice of museum specimens.
- all destructive sampling of material from collections *must* be officially sanctioned by the owner of the material. If destructive sampling is deemed necessary once a loaned specimen is in hand, the researcher must return the specimen to the responsible

curator along with an official request for sampling that indicates the area and amount to be sampled. In the case of herbarium specimens, it may be sufficient to return a photocopy of the specimen, including the label, marked up with the area to be sampled.

- all destructive sampling of herbarium/museum specimens must be reported and acknowledgment given to the owner of the specimen.
- unauthorized use of herbarium/museum material for DNA analysis should be recognized as unethical and actively discouraged by the molecular systematics and herbarium/museum communities.

Recognizing that each request for destructive sampling must be assessed individually on the basis of dialogue between the researcher and the responsible curator, the following are recommended as general principles:

- where large numbers of specimens are required, the researcher must be prepared to travel to the collection to select the material they wish to sample.
- authorization to destructively sample herbarium or zoological material shall be made on a specimen by specimen basis by the curator responsible for the material. This decision to authorize sampling will consider the age and condition of the specimen, its rarity, type status, the amount of material required, and the needs of future researchers. Only in exceptional instances shall the destructive sampling of holotypes be permitted.
- once destructive sampling has been authorized, the sample shall be removed from the specimen in a manner resulting in the least impact on the integrity and scientific value of the specimen (e.g., by using leaf tissue as opposed to floral tissue).
- when ancillary tissue has been placed with a specimen (e.g., material within packets mounted on herbarium sheets, whether specifically prepared for DNA analysis [e.g., air-dried] or not), this material shall be sampled first, provided it can be established that the material has been taken from the same individual organism as the specimen itself.

Research enhances the value of museum voucher specimens. There-fore, it is critical that an association between the specimen and all methodologies and data be maintained in perpetuity. To ensure this, the following procedures are recommended when specimens are de-structively sampled for molecular systematic analyses:

- herbaria/museums require that all specimens used must be an-notated, including records of failed analyses.
- herbaria/museums require that GenBank accession numbers are included on all annotation labels.
- journals require that all molecular systematic treatments are supported by voucher material deposited in recognized her-baria/museum collections. Where appropriate, the material should be accompanied by relevant field or lab notes docu-menting techniques.
- journals require that both GenBank and herbarium/museum voucher numbers are included in all published papers.

Herbaria and other museum collections will play a key role in ad-vancing the frontiers of molecular systematics research by ensuring that:

- new herbarium/museum labels and corresponding databases document information on the preparation and fumigation his-tory of the specimen.
- where appropriate, and as a proactive measure, additional ma-terial is collected and preserved specifically for molecular re-search, e.g. by air-drying or desiccant-drying of plant material.
- all ancillary collection material is linked by database to a voucher specimen.

Finally, all students of plant and animal systematics should be trained in the value and use of herbarium/museum collections, in-cluding accepted etiquette and protocols, and in the broader principles and techniques of collecting and using specimens for all aspects of sys-tematics and for all kinds of systematic investigations.

ACKNOWLEDGMENTS

I thank all the symposium contributors as well as S.C. Byers, T.A. Dickinson, and T. Sage for critical comment on the final recommendations. This is contribution number 55 from the Centre for Biodiversity and Conservation Biology, Royal Ontario Museum.

LITERATURE CITED

ENGSTROM, M.D., R.W. MURPHY, AND O. HADDRATH. 1999. Sampling vertebrate collections for molecular research: practice and policies. Pp. 315-330 in D. A. Metsger and S. C. Byers, editors. *Managing the Modern Herbarium.* Society for the Preservation of Natural History Collections, Washington, DC, xxii+384 pp.

JANSEN, R. K., D.J. LOOKERMAN, AND H-G. KIM. 1999. DNA sampling from herbarium material: a current perspective. Pp. 277-286 in D. A. Metsger and S. C. Byers, editors. *Managing the Modern Herbarium.* Society for the Preservation of Natural History Collections, Washington, DC, xxii+384 pp.

MILLER, J. S. 1999. Banking desiccated leaf material as a resource for molecular phylogenetics. Pp. 331-344 in D. A. Metsger and S. C. Byers, editors. *Managing the Modern Herbarium.* Society for the Preservation of Natural History Collections, Washington, DC, xxii+384 pp.

MUELLER, G.M. 1999. A new challenge for mycological herbaria: destructive sampling of specimens for molecular data. Pp. 287-300 in D. A. Metsger and S. C. Byers, editors. *Managing the Modern Herbarium.* Society for the Preservation of Natural History Collections, Washington, DC, xxii+384 pp.

WHITFIELD, J. B. 1999. Destructive sampling and information management in molecular systematic research: an entomological perspective. Pp. 301-314 in D. A. Metsger and S. C. Byers, editors. *Managing the Modern Herbarium.* Society for the Preservation of Natural History Collections, Washington, DC, xxii+384 pp.

WOOD, E. W., T. ERIKSSON, AND M. J. DONOGHUE. 1999. Guidelines for the use of herbarium materials in molecular research. Pp. 265-276 in D. A. Metsger and S. C. Byers, editors. *Managing the Modern Herbarium.* Society for the Preservation of Natural History Collections, Washington, DC, xxii+384 pp.

351

PART III

ABSTRACTS AND SHORT PAPERS FROM

THE HERBARIUM INFORMATION BAZAAR

Legacy of Mercuric Chloride

LINDA RADER AND CELIA ISON

All herbaria must deal with pest management issues on a daily basis, and some of the techniques employed in the past to prevent damage to plant specimens have created ongoing health hazards for workers in the collections. One method that has potential for serious health problems involves poisoning specimens with a solution of mercuric chloride. During earlier decades of the 20th century, this practice was used in the C.E. Bessey Herbarium (NEB) at the University of Nebraska State Museum in Lincoln. Considering the long tradition of specimen loans, exchanges and mergers of entire collections in herbaria, the issue of mercury contamination is everyone's concern. Steps can be taken to address the problem while still maintaining access to the collection. These include detecting the presence of mercuric chloride in the collection, determining if levels of mercury vapor are dangerous, and establishing workable methods to protect staff and visitors from exposure.

Visible evidence of previous mercuric chloride treatment is often, but not always, presented as blackened mounting paper, "tide lines" of plant pigments dissolved in alcohol that migrate with the solution on the sheet, or crystals deposited on the plant itself. Mercury vapor can off-gas from the sheets in the cases for many years following treatment. Lab safety supply houses sell instruments and chemical products that detect, monitor levels, and absorb mercury vapor in the cases. To minimize personal contact with the mercury salts and vapor, the herbarium staff at NEB wear latex or vinyl gloves while handling specimens and routinely leave case doors open to vent the gas for several minutes prior to working in them. Adequate room ventilation is imperative. Informational sheets on the dangers of mercury poisoning are posted at all times and visitors are verbally advised of necessary precautions.

Coping with the mercury problem has significantly changed handling procedures in the collection at NEB. The measures are obviously cumbersome and create additional collection management problems. Considering the potential for irreversible nervous system damage, however, it seems only prudent to make the effort.

Processing Delicate Aquatic Vascular Plants

THOMAS F. WIEBOLDT AND S. LLYN SHARP

Note: The following procedure is modified slightly from the technique described by Forman and Bridson (1989) and is intended only to illustrate the method rather than present a new one.

INTRODUCTION

Submerged aquatic vascular plants and other delicate aquatic angiosperms present a number of difficulties to mounting by traditional methods using water-based glues. Their extremely thin texture allows them to droop, making it difficult to place them on a glue plate. Removal and transfer to the mounting sheet can be even more problematic. Their tendency to take up water from the glue may cause rapid leaf curling or other distortions of the specimen. For these reasons, water-based glues should be avoided. We use Archer Adhesive (Carolina Biological Supply), a plastic resin with a xylene and alcohol solvent. Other such products are available elsewhere and are equally appropriate. As these chemicals are hazardous, it is imperative that the procedure depicted here be carried out in a fume hood. Once dry, the resin is non-hazardous.

PROCEDURE

First, the label is affixed to the mounting sheet in the usual manner and the specimen carefully positioned on the sheet. If a fragment packet is part of the collection, this should only be attached to the sheet after the specimen has been mounted.

Next, a sheet of waxed paper is laid over the specimen aligning one edge and corner with the mounting sheet. To facilitate handling, a stiff cardboard is used both below the mounting sheet and on top of the waxed paper so that the entire block (bottom cardboard, mounting sheet, specimen, waxed paper, top cardboard) can be inverted. Then, being careful not to disturb the specimen, the upper cardboard and mounting sheet are removed to expose the specimen which is now upside down on the waxed paper.

Gluing is accomplished by holding the specimen in place and dispensing the glue from a plastic squeeze bottle, spotting it along the stems and leaves of the specimen. (No precise details can be given as to where and how much glue to use. The idea is simply to attach it to the mounting sheet so that no long sections are free to move about. Generally, it is better to glue stems than leaves as these are structurally stronger.)

Next, the mounting sheet is laid face down on top of the specimen, care being taken to align its edge and corner with the waxed paper as noted previously. This assures that the label will be positioned properly relative to the specimen. The sheet is pressed into the glue by gently hand-rubbing over the back.

Then, placing the cardboard back on top, the block is again turned, returning it to the upright position. The top cardboard is removed and the waxed paper slowly peeled back to expose the specimen. Additional glue may be spotted on the top surface if necessary or desirable.

Small metal washers can be used to weight the specimen and keep it firmly in the glue. It works well to encircle the glue spot with the washer. Small ceramic tiles can be used during drying as spacers between specimens, forming a small stack. Tiles are placed in spaces between plants or plant parts, the positions varying depending on the layout of the specimen on the mounting sheet.

ADDITIONAL NOTES

A technique often mentioned in regard to aquatic plants is that of floating the specimen directly onto the mounting sheet in a shallow tray. By slowly lifting the sheet out of the water, the specimen is stranded in a fairly natural pose making a nice-looking mount. The mounting sheet can then be placed in the press and dried (with a sheet of waxed paper on top of the specimen to prevent it adhering to the papers or blotters). The disadvantage of this method is that, while the specimen at first seems fairly well attached to the sheet, it will eventually become detached since no adhesive has been used. Strapping can, of course, be added, but the specimen will still be relatively loose.

Aquatic specimens pressed in newsprint in the usual way often become stuck to the newsprint. Removal invariably breaks up the specimens and often leaves unsightly spots of newspaper attached. To prevent sticking, waxed paper or other paper less absorbent than newsprint can be used either singly or both above and below the specimen in pressing. Although waxed paper might not seem porous enough, it will in fact absorb water to a certain extent and will have no deleterious effect, especially if a drying oven is used or newsprint and blotters are changed often to more effectively remove the moisture.

Delicate aquatic specimens, mounted by even the most effective means, are extremely fragile. Long-term stability can be insured only if this is kept in mind and great care in handling is always practiced.

LITERATURE CITED

FORMAN, L. AND BRIDSON, D. 1989. *The Herbarium Handbook*. Royal Botanic Gardens, Kew.

Methyl Cellulose for Mounting Plant Specimens at the San Diego Natural History Museum Herbarium

JUDY GIBSON

The herbarium staff at San Diego Natural History Museum (SD) began trial use of methyl cellulose (MC) in 1991 after curator Geoff Levin read a favorable report on its archival qualities (Clark, 1986). Late that year SD adopted this material as its primary adhesive for mounting of dried herbarium specimens. Our overall assessment is that, while it is inferior to poly(vinyl acetate) (PVA) white glues in certain respects (noted below), it has proved satisfactory in all essential respects in general herbarium use.

ADVANTAGES OF MC COMPARED TO PVA

- Nontoxic
- Easy to clean smears off the sheet
- Can be mixed in small quantities
- Reversible with water
- Does not become acidic with age
- Does not thicken or dry around the edges as you work
- Easy to clean up work area and tools

DISADVANTAGES OF MC COMPARED TO PVA

- Makes a weaker bond

- requires firm strapping
- Not very *sticky*: can't be sure things are glued until they dry
- If mixed too thick specimens get stuck in it (using glass plate method)
- Takes longer to dry
- High water content contributes to wrinkling and curling of labels
- Does not bridge gap between specimen and paper
- Cannot be thinned at will to suit various types of plant material

MIXING

Methyl cellulose powder will dissolve only in warm water and thickens only when it cools. We use a strength of about 2.5–3% (specifically, 1 ½ tablespoons powder for a 230 ml jar of adhesive). The powder weighs about 3.7 g per tablespoon.

Mix the powder thoroughly in about one third of the needed quantity of hot (near boiling) water, then quickly stir in the remaining two-thirds of the water, very cold. The mixture will thicken instantly. If you mix the powder in the full quantity of hot water it will tend to settle out before it cools, though it can be stirred after cooling.

The exact proportions are not critical. A mixture half as thick as above will be so weak that labels can be peeled off the sheet when dry. A mixture twice as thick will be too gummy to use. After mounting about 5000 specimens, we had not yet finished our first (one-pound) jar of powder.

MOUNTING PLANTS

We apply the adhesive either by the glass plate method or, for delicate or wispy specimens, section by section with a thin knife blade. We also use it at the same strength for labels and packets. We keep PVA on hand for bulky objects like nuts or pieces of bark as well as for coriaceous, smooth or waxy leaves, for which methyl cellulose is inadequate.

This adhesive takes longer to dry than PVA. We find that specimens, and especially the labels, are not completely dry until at least 24 hours after mounting. We place a stiff card over the labels to hold them flat until they are removed from the pressing box. If removed too soon, the labels curl.

The specimens are reinforced with linen straps to prevent their being pried loose from the sheet. Because of the weakness of the bond we use more and wider straps than before. As the straps are added, the specimen is examined for portions that are not well-adhered to the sheet. Adhesive is applied with a knife blade if needed.

REMOVING OBJECTS FROM THE SHEET

Sturdy specimens can often be removed from the sheet by carefully prying up each section. It might be necessary to moisten leaves around their edges. If a specimen will tolerate being wetted, it might be possible to spray it with water (we have not tried this). For more delicate specimens, cover the plant with damp paper towels until it is moistened. Gently lift each portion with fine tools (e.g., dental tools, thin knife blade). Labels can also be moistened and removed.

SOURCES

These suppliers sell one-pound jars of methyl cellulose powder from City Chemical Co. of Jersey City, New Jersey:

- University Products, PO Box 101, Holyoke, MA 01041-0101, USA; (413)532-9431 or (800)762-1165
- Pacific Papers, PO Box 606, Cotati, CA 94931, USA; (800)676-1151

- Herbarium Supply Company 3483 Edison Way, Meulo Park, CA 94025-1813, USA; (800) 348-2338 (415) 366-8868

Literature Cited

Clark, S. H. 1986. Preservation of herbarium specimens: An archive conservator's approach. Taxon 35(4):675-682.

[Editors' note: The San Diego Natural History Museum herbarium has recently discontinued the use of methyl cellulose due to the weakness of the adhesive bond.]

A New Technique for Drying and Mounting Spruce and Hemlock Specimens

DEBORAH LEWIS AND RUTH HERZBERG

Herbarium specimens of some of the conifers, especially spruce (*Picea*) and hemlock (*Thuja*), have not turned out well using traditional drying techniques. When specimens of these genera are dried in a plant press, delays in processing the specimens combined with drying over heat tend to cause most or all of the needles to drop. A new approach to drying conifer specimens has been developed to try to reduce needle loss and so improve the quality of the specimens.

A laminating/dry mount press is used to rapidly flatten the specimens and initiate the drying process. The temperature of and length of time in the laminating press varies between taxa: hemlock usually requires a shorter time and lower temperature than spruce. Once flattened, the specimen is mounted while it is still damp. The standard technique for mounting is 1) dipping it in glue and affixing it to a sheet, 2) covering it with a sheet of waxed paper and protective corrugate, and 3) applying weight to the specimen and allowing it to air dry. For optimum results the specimens should be processed as soon as possible after collecting, or, if there is a delay, should be kept cool and moist to minimize drying before the specimen is processed. This new approach yields a net reduction in needle loss on specimens of these two genera.

A Protective Hardboard Folder for Storing Valuable Herbarium Specimens

Michael J. Shchepanek

[Editors' note: This was originally a demonstration.]

Most herbaria store their valuable specimens, such as types, mounted on herbarium paper, within closed or semi-closed soft cardboard folders. In handling or shipping these folders, the specimens are susceptible to damage. To alleviate this problem, an archival hardboard folder (Fig. 1) was designed and constructed to provide maximum protection to valuable specimens both in storage and in handling. The design is versatile so that herbarium sheets of different dimensions and specimens of various sizes and thicknesses can be accommodated. These folders can be easily made from materials available from art supply stores.

Figure 1. An archival hardboard folder for storing valuable herbarium specimens.

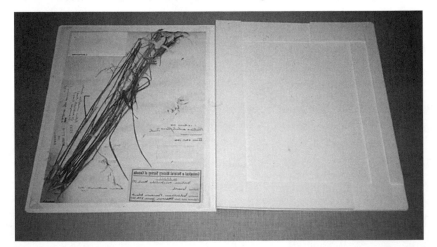

Shipping Herbarium Specimens:
The Egg Carton Solution

ANN PINZL

Packing herbarium specimens for shipping often requires some sort of extra filler or padding to be placed in the shipping container along with the specimens. Cartons in which a dozen eggs are purchased can be used as filler and/or shock absorber. Their approximate size of 11½ x 4 ins. (29.2 x 10.2 cm) meshes nicely with herbarium shipment parcels; four egg cartons, placed next to each other along their long side, neatly covers the cardboard ventilators usually used to pack plant specimens. The egg cartons are easy to acquire, use, and remove (no flying styrofoam pellets). Further, they are reusable and are low (no) cost.

Retrofitting of Collections
Storage Case Gaskets

Karen L. Johnson and Janis Klapecki

The Manitoba Museum of Man and Nature houses all of its natural history specimens in one large collections storage area. In the past, a number of serious insect infestations occurred in both the animal and plant collections due to problems with environmental control in this area. Deteriorating door seals on a number of older herbarium and other specimen cases added to the problem. The seals were made of materials such as felt, foam, and natural rubber. The felt, which dermestid beetles and moths can and do use as food, was ancient (Fig. 1); the foam was disintegrating; and the natural rubber had become greatly compressed and no longer provided a proper seal.

A complete retrofit of the collections storage area in the early 1990s allowed the Museum's Botany Department to replace damaged and inappropriate seals on all herbarium cases in 1993.

MATERIALS AND METHODS

Polyethylene sheeting was taped in place to cover exposed herbarium specimens with the case door left ajar. The sheeting helped to keep debris from entering the specimen compartments and

Figure 1. Felt gasket (adhesive unknown).

Figure 2. Newly installed silicone rubber gasket.

maintained a more stable internal case environment for the period during which they were to remain open. Existing gaskets were physically removed using putty knives and any remnants of adhesive were dissolved with acetone.

The Canadian Conservation Institute (CCI) recommended plain unbacked silicone rubber gaskets (Fig. 2) adhered with GE Silicone II (neutral curing). Three thicknesses (1/8 in., 3/16 in., and 3/8 in.[3 mm, 4.8 mm, and 9.5 mm]) of gasket were used to match the different gaps in the several types of cases retrofitted (herbarium, osteology, and over-sized metal cases). All thicknesses of gasket came in three-foot lengths and widths of 9/16 in. (15 mm). Gaskets were supplied by: E. F. Walter Ltd., 51 Wingold Ave., Toronto, ON M6B 1P8; (416) 782-4492 (telephone), (416) 782-2190 (fax).

As GE II Silicone was not available in Winnipeg at that time, GE SilglazeN 2500 was suggested by the distributer to be similar to GE II Silicone and was subsequently used for this project. Although the manufacturer stated that SilglazeN 2500 was neutral curing, a small amount of acetic acid fumes was, in fact, given off during the curing process. This odour dispersed rapidly (within 24 hours) and we did not perceive it to cause any problems.

The silicone rubber gasket was measured and cut to fit each case. Corners were butt-jointed rather than mitred (Raphael and Cumberland, Jr., 1992) as this was a far easier cut and joint to make and did not seem to jeopardize the effectiveness of the joint seal. A thin bead of the silicone adhesive was then run continuously around the perimeter of the case. Each join and corner joint in the gasket was filled with the adhesive to prevent any breaks in the seal. The silicone adhesive was allowed to cure for two to three days with the case doors open to aid in the dispersion of gasses. All of the cases were in the closed collections storage area and any fumes given off were exhausted through the internal environmental control system.

Cost for the 54 specimen cases retrofitted was between CAN$25 and $30 per case. Labour (mostly paid from a federal government grant) ran between two and three hours per case depending on the size of the case and the type of old gasket which had to be removed.

RESULTS

Evaluating the retrofit after two years, we were very pleased with the results. The silicone rubber gaskets had stood up very well, with only slight signs of compression where the doors close and no signs of lift in the silicone adhesive. The slight compression observed may be due to the *very* tight fit of the gaskets as the closest available thickness of silicone rubber was used to fill the gap. Insect infestations have been non-existent to date because of the new seals and vigilant monitoring (we use no chemical fumigants).

NOTE

Since the completion of this project, Jean Tétreault, Conservation Scientist at CCI, was consulted on the use of the SilglazeN 2500 product. Mr. Tétreault suggested that the product was probably listed as "neutral curing" because it releases much lower amounts of acetic acid in relation to other similar sealants. The resultant off-gassing from this product ceased within two to three weeks and should have no lingering detrimental effects on the specimens or paper mounts stored in these cases. Visual inspection of specimens concurs with this. Mr. Tétreault recommends two new products from Dow Corning that are indeed neutral curing (i.e., do not release acetic acid or ammonia) and are recommended over the GE II silicone product. These products are Dow Corning 737 and Dow Corning Silastic Q3-3744 RTV. Even though these two new products do not release acetic acid, Mr. Tétreault still recommends that the cases be allowed to remain open for a period of two to three weeks to allow the solvent in the silicone adhesive to evaporate.

LITERATURE CITED

RAPHAEL, T. AND D. R. CUMBERLAND, JR. 1992. Retrofitting steel storage cases: installing new, improved gasketry. Pp. 233-234 in C. L. Rose and A. R. de Torres, editors. *Storage of Natural History Collections: Ideas and Practical Solutions.* Society for the Preservation of Natural History Collections, Iowa City, IA.

Comments on Freeze-Drying and Storing Entire Mycological Specimens

SIMON MOORE

The county of Hampshire, in the mid-south of England, is fortunate in having a diversity of ideal habitats that produce a wide-ranging mycota. The New Forest alone comprises heathland, grassland, acid bog, birch, beech, and other deciduous woodland, and also areas of pine providing a wide and varied succession of mycorrhizal associations. In addition to the usual "fungus kickers," the recent trend toward extending gustatory appreciation of wild fungi by restaurateurs has caused some depredations to the mycota in certain areas of the Forest. Because of this, the Hampshire County Council Museums Service, in collaboration with the New Forest Museum and a group of local mycology experts, are setting up a mycology herbarium to preserve specimens of as many species as possible before they become rare or even disappear from the area.

Although presently the collection only runs to some 300 species, it is hoped that local interest and enthusiasm will help to augment this number gradually over the years and that in time a scientifically-useful reference collection will result. To attract attention to this project many whole specimens of the more eye-catching species have already been freeze-dried and featured in local museum dioramas.

PROCESSING

The specimens are taken, with full data, and freeze-dried in an old 1970s Edwards EF2 freeze-drier. The specimens are processed at −30°C and at a vacuum decreasing to just below 0.1 atmospheres (50 Torr); the slightly lower temperature helps to prevent rapid lyophilization taking place which leads to cracking of the caps and general distortion of fragile specimens. This process normally results in morphologically-perfect specimens, although some lightening of sensitive pigments such as muscapurpurin and muscarubrin, found in the cap of the fly agaric (*Amanita muscaria*), does occur as the process appears to reduce the wavelength of these pigments by up to 100 nm (Dopp et al., 1971). In the fly agaric, this may also be due to the conversion through drying of the hallucinogen ibotenic acid, found in the skin of the cap, to muscimol. Fungal smells, which help in identification, are occasionally, although temporarily, preserved by freeze-drying. However, the strong curry smell of *Lactarius camphoratus* is also preserved by simple air drying.

STORAGE

Freeze-dried specimens are presently stored in a building approximately 150 years old. When the herbarium was started in 1993, conditions in the building in winter averaged about 5°C and 85% relative humidity (RH). When one dehumidifier was installed about 30 metres from the mycological collection, the RH in the storage area ranged between 50% and 70%. With the addition in 1994 of a second dehumidifier closer to the collection, the temperature has increased to about 10°C to 15°C and the RH has fallen gradually to 45%. The specimens are stored in seal-top poly bags with a minimum amount of air inside but enough to prevent the bags from crushing or distorting the specimens. Silica gel is not placed inside the storage bags since this would cause physical damage to the specimens. The bags themselves are stored in a pine drawer unit with clip-on doors to reduce the bleaching effect of natural and artificial light.

DETERIORATION

When the RH rose to above 55% a slight crinkling of the caps of many basidiomycete specimens was observed. If unchecked this gradually caused the entire cap to wrinkle; whole specimens became soft and pliable although those with fugitive/sensitive pigmentation showed no signs of further fading. Ultimately, affected specimens would have deliquesced had the second dehumidifier not been installed.

INSECT PESTS

Some beetle larvae of *Cis bilamellatus* actually survived the freeze-drying process and metamorphosed inside a birch polypore (*Piptoporus betulinus*). Fortunately, before any eggs could be laid, specimens were re-frozen at −20°C and the adults were killed. Fungal specimens displayed in dioramas are still susceptible to the depredations of museum beetles and moths although spraying with a varnish (high gloss for specimens with a viscid appearance, semi-matt for others) does seem to render the specimens unpalatable to insect pests.

PREPARATION OF HERBARIUM SPECIMENS

Freeze-dried fungi are now being prepared for a myco-herbarium. Fruiting bodies are freeze-dried and then sliced through the midline; the median slice is stored in an acid-free paper folder mounted onto a herbarium sheet. The slice can then be examined microscopically to view the arrangement and types of cystidia and any other morphological characters. Examination of slices has to be carried out in a medium-dry atmosphere since the slices are sensitive to an RH greater than 55% and some have already shown that they can distort and shrink up to 50% if left overnight in a room with an RH of only 60%. Where possible, spore prints are taken and some spores mounted onto microslides.

COMMENTS ABOUT FREEZE-DRYING FUNGI

The advantage of freeze-drying fungi for use in a herbarium is that shrinkage and colour loss are reduced to 0% to 5%. However, curators and preparators should be aware of the hygroscopic properties of freeze-dried fungi mentioned above. Further, freeze-dried gill fungi are fragile and liable to fracturing, particularly the gill area (e.g., *Lepiota* spp.) and the stem if hollow (e.g., *Coprinus* spp., *Hygrocybe* spp.). Breakages should be repaired using a solvent-based glue; water-based adhesives, such as PVA, cause partial rehydration of the repaired tissue leading to shrinkage, distortion, and discoloration.

LITERATURE CITED

DOPP, H., W. GROB, AND H. MUSSO. 1971. *Uber die Farbstoffe des Fliegenpilzes.* Naturwissenchaften 58(11):566.

INDEX

O R D E R   F O R M

MANAGING THE MODERN HERBARIUM
An inter-disciplinary approach
D.A. Metsger and S.C. Byers, Editors

RŎM

S P N H C

Published by
Society for the Preservation of Natural History Collections
Washington, DC

CANADIAN FUNDS (FOR SHIPMENT IN **CANADA** ONLY)		U.S. FUNDS (FOR SHIPMENT TO **U.S.** AND **OVERSEAS**)	
__ Copies @$39.95	$_____	__ Copies @$29.95us	$_____
GST (7%)	$_____	Shipping (1st book) within US	$ 5.75us
Shipping (1st book)	$ 6.00	Add $3.00us for each additional book	$_____
Add $3.00 for each additional book	$_____	Shipping (1st book) OVERSEAS	$ 6.50us
		Add $3.25us for each additional book	$_____
Total enclosed	$_____	**Total enclosed**	$_____

Make cheque or money order payable to:

ELTON-WOLF PUBLISHING
212 - 1656 Duranleau Street
Vancouver, BC Canada V6H 3S4
Telephone: (604) 688-0320 • Fax: (604) 688-0132

Name _____

Address _____

City, Province/State _____

Postal/Zip Code_____

Phone (work) _____ (home) _____

Thank you for your order!